Rare Metal Technology 2016

TMS2016

145th Annual Meeting & Exhibition

FEBRUARY 14-18 DOWNTOWN NASHVILLE, TENNESSEE MUSIC CITY CENTER

New proceedings volumes from the TMS2016 Annual Meeting:

- 7th International Symposium on High-Temperature Metallurgical Processing
- CFD Modeling and Simulation in Materials Processing 2016
- Characterization of Minerals, Metals, and Materials 2016
- Energy Technology 2016: Carbon Dioxide Management and Other Technologies
- EPD Congress 2016
- Light Metals 2016
- Magnesium Technology 2016
- Rare Metal Technology 2016
- REWAS 2016
- Shape Casting: 6th International Symposium
- TMS 2016 Supplemental Proceedings

Rare Metal Technology 2016

Proceedings of a symposium sponsored by
the Hydrometallurgy and Electrometallurgy Committee
and the Extraction & Processing Division of
The Minerals, Metals & Materials Society (TMS)

held during

TMS2016
145th Annual Meeting & Exhibition
FEBRUARY 14-18 DOWNTOWN NASHVILLE, TENNESSEE MUSIC CITY CENTER

Edited by:

Shafiq Alam, Hojong Kim, Neale R. Neelameggham, Takanari Ouchi, and Harald Oosterhof

Editors
Shafiq Alam
Hojong Kim
Neale R. Neelameggham
Takanari Ouchi
Harald Oosterhof

ISBN 978-3-319-48616-1
ISBN 978-3-319-48135-7 (eBook)
DOI 10.1007/978-3-319-48135-7

Chemistry and Materials Science: Professional

Published by Springer International Publishers, Switzerland, 2016
Reprint of the original edition published by John Wiley & Sons, Inc., 2016, 978-1-119-23107-3

Printed on acid-free paper

This Springer imprint is published by Springer Nature
The registered company is Springer International Publishing AG
The registered company address is: Gewerbestrasse 11, 6330 Cham, Switzerland

TABLE OF CONTENTS
Rare Metal Technology 2016

Rare Earth Elements / Base & Rare Metals I

Rare Earth Elements / Base & Rare Metals II

Platinum Group Metals / Mo, Ti, V & W

PREFACE

Rare Metal Technology 2016 is the proceedings of the symposium on Rare Metal Extraction & Processing sponsored by the Hydrometallurgy and Electrometallurgy Committee of the TMS Extraction and Processing Division. The symposium has been organized to encompass the extraction of rare metals as well as rare extraction processing techniques used in metal production. This is the third in a series of symposia, which started in 2014. The symposia have been held in San Diego, California; Orlando, Florida; and now in Nashville, Tennessee.

The intent of the symposium was to avoid conflicts with major international symposia during 2015–2016 such as International Hydrometallurgy Conference, International Precious Metals Conference, etc., while covering the extraction of rare/less common metals or minor metals (not covered by other TMS symposia). The elements considered included antimony, bismuth, barium, beryllium, boron, calcium, chromium, gallium, germanium, hafnium, indium, manganese, molybdenum, platinum group metals, rare earth metals, rhenium, scandium, selenium, sodium, strontium, tantalum, tellurium, tungsten, etc. These are rare metals of low tonnage sales compared to high tonnage metals such as iron, copper, nickel, lead, tin, zinc, or light metals such as aluminum, magnesium, or titanium and electronic metalloid silicon. Rare processing included bio-metallurgy, hydro-metallurgy, electro-metallurgy, as well as extraction of values from electric arc furnace (EAF) dusts, and less common waste streams not discussed in recycling symposia. Rare high-temperature processes included microwave heating, solar-thermal reaction synthesis, and cold crucible synthesis of the rare metals and the design of extraction equipment used in these processes as well as laboratory and pilot plant studies.

This proceedings volume covers many rare metal and rare-earth elements where papers are presented on the extraction and processing of elements like antimony, arsenic, gold, indium, palladium, platinum, rare earth metals including yttrium and neodymium, titanium, tungsten, vanadium, etc. Uses of biomass wastes in extractive metallurgy befit the rare processing title. Rare processing techniques includes direct extraction process for rare earth element recovery, biosorption of precious metals, fluorination behavior of uranium and zirconium mixture of fuel debris treatment, recovery of valuable components of commodity metals such as zinc, nickel, and metals from slag. The symposium is organized into sessions encompassing (1) Rare Earth Elements, (2) Base and Precious Metals, (3) Platinum Group Metals, and (4) Molybdenum, Titanium, Vanadium & Tungsten.

We acknowledge the efforts by the organizing and editing team consisting of Shafiq Alam, Hojong Kim, Neale R. Neelameggham, Takanari Ouchi, and Harald Oosterhof. Our thanks to Trudi Dunlap and Patricia Warren of TMS in assembling the proceedings, and Matt Baker for facilitating the proceedings. We sincerely thank all the authors, speakers, and participants and look forward to continued collaboration in the advancement of science and technology in the area of rare metal extraction and processing.

Shafiq Alam
Lead Organizer

EDITORS

Shafiq Alam is an associate professor at the University of Saskatchewan, Canada. In 1998, he received his Ph.D. degree in chemical engineering from Saga University, Japan. From 1999–2001, he was appointed as a post-doctoral research fellow at the University of British Columbia and the University of Toronto, Canada.

Dr. Alam has extensive experience in industrial operations, management, engineering, design, consulting, teaching, research, and professional services. Before joining the University of Saskatchewan in 2014, he was an assistant/associate professor at Memorial University of Newfoundland for about seven years. Prior to starting his career in academia, he worked with many different companies, such as, Shell, Process Research ORTECH Inc., Fluor Canada Ltd., and the National Institute of Advanced Industrial Science and Technology (AIST), Japan. Dr. Alam is highly experienced in the area of mineral processing and extractive metallurgy, and he possesses two patents and has more than 134 publications. He is the co-editor of three books and an associate editor of the *International Journal of Mining, Materials and Metallurgical Engineering* (*IJMMME*). He is the winner of the 2014 TMS Extraction & Processing Division's Technology Award.

Dr. Alam is a registered professional engineer and has worked on projects with many different mining companies including, Falconbridge, INCO (Vale), Barrick, Hatch, Phelps Dodge, Rambler, Anaconda, etc. He is an executive committee member of the Hydrometallurgy Section of the Canadian Institute of Mining, Metallurgy and Petroleum (CIM) and currently, he holds the office of "Secretary" (2013–2016). Dr. Alam is also the chair of the Hydrometallurgy and Electrometallurgy Committee of the Extraction and Processing Division (EPD) of The Minerals, Metals & Materials Society (TMS) for the period of 2015–2017. He is a co-organizer of many symposia at the international conferences through CIM and TMS. Dr. Alam is the lead organizer of the present Rare Metal Extraction and Processing Symposium at the TMS 2016 Annual Meeting & Exhibition.

Hojong Kim is an assistant professor in Materials Science and Engineering at the Pennsylvania State University. Dr. Kim received his B.S from Seoul National University in South Korea in 2000 and Ph.D. degree at MIT in 2004 both in Materials Science and Engineering. His doctoral research sought to identify the corrosion mechanisms of constructional alloys in high temperature and high pressure steam environments in the Uhlig Corrosion Laboratory at MIT. After graduate research, Dr. Kim worked as a senior research engineer at Samsung-Corning Precision Glass Co. Ltd. to improve the process yield for thin film transistor liquid crystal display (TFT-LCD) glass manufacturing by engineering high temperature refractory materials. After five years of industrial experience, Dr. Kim returned to MIT as a post-doctoral associate and later as a research scientist to contribute to the growing need for sustainable technology. He conducted research on high temperature electrochemical processes, including molten oxide electrolysis for carbon-free iron production and liquid metal batteries for large-scale energy storage. His current research interests at Penn State include electrochemical separation of materials in molten salt electrolytes, high temperature corrosion, and materials for energy storage and conversion.

Neale R. Neelameggham is "The Guru" at IND LLC, involved in technology marketing and international consulting in the field of light metals and associated chemicals (boron, magnesium, titanium, lithium and alkali metals), rare earth elements, battery and energy technologies, etc. He was a visiting expert at Beihang University of Aeronautics and Astronautics, Beijing, China, and a plenary speaker at the Light Metal Symposium in South Africa – on low carbon dioxide emission processes for magnesium.

He has more than 38 years of expertise in magnesium production and was involved in process development of its startup company NL Magnesium through to the present US Magnesium LLC, UT from where he retired in 2011. He is developing thiometallurgical processes—a new concept of using sulfur as the reductant and/or fuel. He has authored a 21st century convective heat transfer model of continued global anthropogenic warming moderated by excess precipitation-based on thermal emissions independent of energy conversion source–fuel or renewable sources.

Dr. Neelameggham holds 16 patents and applications. He has served in the Magnesium Committee of the TMS Light Metals Division (LMD) since its inception in 2000, chaired the committee in 2005, and in 2007 he was made a permanent co-organizer for the Magnesium Technology Symposium. He has been a member of the Reactive Metals Committee, Recycling Committee, Titanium Committee, and Program Committee Representative of LMD and the LMD Council.

Dr. Neelameggham was the inaugural chair, when in 2008, LMD and EPD (Extraction & Processing Division) created the Energy Committee, and has been a co-organizer of the Energy Technology symposium and co-editor of the proceedings through the present. He received the LMD Distinguished Service Award in 2010. While he was the chair of Hydro and Electrometallurgy Committee he initiated the Rare Metal Technology Symposium in 2014. He is co-editor for the symposia proceedings *Magnesium Technology 2016*, *Energy Technology 2016*, *Rare Metal Technology 2016*, and the light metals section of *REWAS 2016*.

Takanari Ouchi is a research scientist in the Materials Processing Center at Massachusetts Institute of Technology. He received his B.S. in Applied Chemistry (2007), and his M.S. (2008) and Ph.D. (2011) in Nano-Science and Nano-Engineering all from Waseda University. He developed electrochemical deposition processes to fabricate metal nano-structures with both well-controlled crystallinity and uniformity at the single nano-meter scale and demonstrated the applicability of these processes to fabrication of bit patterned magnetic recording media for future hard disk drives. He taught in the Department of Applied Chemistry at Waseda University as a research associate in 2009–2011. After completing his doctoral degree, Dr. Ouchi joined MIT as a postdoctoral researcher. He has developed liquid metal batteries, which is, in principle, a bi-directional electro-refining cell, to apply for the grid scale energy storage. He developed novel calcium and sodium liquid metal batteries by mitigating their solubility in molten salt electrolytes, which is a lasting issue in electrowinning and electrorefining processes. He has taught several undergraduate and graduate students at MIT. Dr. Ouchi is an author of 14 peer reviewed papers and of 46 conference presentations, and has earned several awards and honors, such as TMS EPD Young Leaders Professional Development Award in 2015. He is currently a guest editor of *JOM*.

Harald Oosterhof graduated as a chemical engineer from Twente University (TU) in The Netherlands in 1994. In the same year, he assumed a position as researcher at TU Delft where he worked in the laboratory for process equipment. His research on anti-solvent crystallization of well-soluble salts was rewarded with two patents and several publications. After receiving his Ph.D. from Delft University in 1999, he assumed the position of project manager at Umicore, a global materials and technology group that is based in Belgium. During his first assignment as Project Leader – Hydrometallurgy, he focused on the refining of cobalt, nickel, and germanium. Since 2011, Dr. Oosterhof has worked as scientist in the Recycling and Extraction Technology group at Umicore's Central R&D Department. His main competence areas are special metals hydrometallurgy, recycling and refining of rare earth metals, base metal hydrometallurgy, and recycling of spent rechargeable batteries and superalloys. In his current position, Dr. Oosterhof is frequently involved in business development of scarce metals recycling and he is heading a team of hydrometallurgical specialists

SESSION CHAIRS

Rare Earth Elements / Base & Rare Metals I

Harald Oosterhof

Takanari Ouchi

Rare Earth Elements / Base & Rare Metals II

Shafiq Alam

Hojong Kim

Platinum Group Metals / Mo, Ti, V & W

Neale R. Neelameggham

Hojong Kim

Rare Metal Technology

2016

Rare Earth Elements / Base & Rare Metals I

Rare Metal Technology 2016
Edited by: Shafiq Alam, Hojong Kim, Neale R. Neelameggham, Takanari Ouchi, and Harald Oosterhof
TMS (The Minerals, Metals & Materials Society), 2016

THE SEARCH MINERALS DIRECT EXTRACTION PROCESS FOR RARE EARTH ELEMENT RECOVERY

David Dreisinger[1], Niels Verbaan[2], Mike Johnson[2]

[1]Search Minerals, #211, 901 West 3rd Street
North Vancouver, BC, Canada V7P 3P9
[2]SGS Minerals Services, PO Box 4300
185 Concession Street, Lakefield, ON, Canada K0L 2H0

Keywords: Rare Earths, Extraction, Mixed Rare Earth Oxide, Foxtrot

ABSTRACT

The Search Minerals Foxtrot project in Labrador is a surface deposit enriched in the highly sought after heavy rare earth elements. Early metallurgical work investigated beneficiation of the rare earth content of the Foxtrot resource using flotation, gravity separation and magnetic separation. The concentrate was then processed by acid baking and water leaching to produce a REE leachate for purification and recovery.

The Search Minerals Direct Extraction Process was developed through bypassing the beneficiation process and directly acid treating a coarse crushed ore material prior to water leaching. The result is a simple, low risk technical process for direct recovery. Concurrently, the treatment of the water leach solution was modified to produce a high grade (98.9% total rare earth oxide) product for refining.

INTRODUCTION

Search Minerals Inc. (Search) is exploring and developing a number of deposits for rare earth element (REE) recovery in Labrador, Canada (Figure 1). The Port Hope Simpson District is approximately 135 km by 12 km in size, and is highly prospective for heavy and light rare earth elements. The Foxtrot deposit sits within the Port Hope Simpson REE District. The infrastructure available at Foxtrot is excellent; a deep water port, air strip, road and power infrastructure are preexisting at Port Hope Simpson (Figure 2). The three communities of Port Hope Simpson, St. Lewis and Mary's Harbour are in close proximity to the site.

The Foxtrot deposit is estimated to have an indicated mineral resource of 9.229 Mt at a grade of 1.07% TREO (0.21 % HREO and 0.85% LREO) and an inferred resource of 5.165 Mt at a grade of 0.93% TREO (0.20 % HREO and 0.73% LREO) at a Dy cut-off grade of 130 ppm (Srivastava et al, 2013). The deposit contains approximate 20% as HREO (Table 1).

The early metallurgical testwork on Foxtrot ore samples focused on mineralogy, beneficiation, concentrate sulphation and water leaching, followed by PLS purification and recovery of rare earths as a mixed oxalate precipitate (Dreisinger et al, 2012). Bastnasite, synchysite, monazite, chevkinite, fergusonite and allanite were identified in the mineralogical work. Allanite and fergusonite carry most of the Dy at 49.3% and 40.5%, respectively, followed by chevkinite (8.8%) and bastnasite (1.4%). Further work was also reported in 2014 (Dreisinger et al).

Figure 1 – Location Map for Search Minerals Exploration and Development in Labrador Canada.

Gravity, magnetic separation and flotation were applied to a bulk sample of Foxtrot ore to produce an upgraded mineral concentrate. The reported recovery averaged 83% to a concentrate containing 38.5% of the original mass. The concentrate was then subjected to acid baking and water leaching to produce an acid leachate containing the rare earth sulphates. The solution was the purified by pH adjustment to reject iron, thorium and some degree of aluminum. Rare earths were subsequently precipitated with oxalic acid to make a mixed rare earth oxalate containing over 55% TREO content.

The key issues and observations with respect to the early metallurgical testing of the Foxtrot resource can be summarized as follows;

1. Complex beneficiation: The use of gravity, magnetic and flotation separation to produce a high mass recovery concentrate with ~17% loss of rare earth values to tailings is costly in capital and operating cost and the loss of rare earth elements is a heavy penalty on potential mine revenue. The use of three separate beneficiation techniques for minimal benefit in overall concentration of value does not appear to be justified.
2. Optimization of Reagent Use: The use of sulphation and water leaching was highly effective at extracting the rare earth elements from the mineral concentrate.

However, there was little opportunity for optimization of the sulphation conditions in the early work.
3. Impurity Control: The use of simple pH adjustment to reject thorium followed by oxalate precipitation was effective in recovering a crude mixed rare earth oxalate. However, the levels of thorium (at 163 ppm in the oxalate precipitate) are likely too high to be acceptable to a rare earth refinery.

The purpose of the current study was to simplify the process flowsheet by attempting to treat the whole ore (to avoid cost and metal loss through beneficiation), reduce the operating cost per unit of recovered rare earth oxide and produce a premium quality product (mixed rare earth oxide) low in thorium for further refining. The study was first conducted with batch testing of the whole ore treatment followed by a second phase where acid-ore mixtures were treated in a continuous kiln followed by batch leaching and purification/recovery steps.

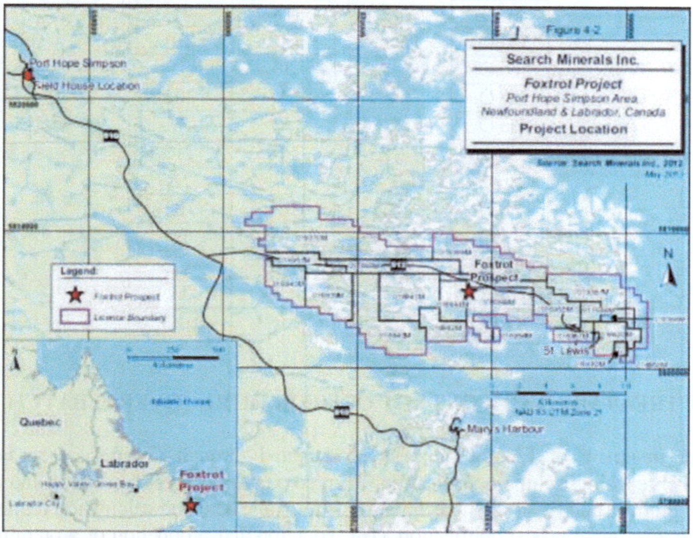

Figure 2 – The Foxtrot Deposit within the Port Hope Simpson REE District in Labrador Canada

Table 1 – Mineral Resource Estimate for the Foxtrot Project (Srivastava et al, 2013). CIM Mineral Resource Definitions followed. Cutoff grade of 130 ppm Dy. LREO = Oxides of La+Ce +Pr+Nd+Sm, HREO = Oxides of Eu,+Gd+Tb+Dy+Ho+Er+Tm+Yb+Lu+Y, TREO = LREO + HREO.

Resource	Units	Indicated	Inferred
Mass	t	9,229,000	5,165,000
Y_2O_3	(ppm)	1320	1237
La_2O_3		1926	1669
CeO_2		4105	3544
Pr_6O_{11}		465	400
Nd_2O_3		1687	1442
Sm_2O_3		303	264
Eu_2O_3		15	13
Gd_2O_3		217	210
Tb_4O_7		38	35
Dy_2O_3		217	202
Ho_2O_3		42	39
Er_2O_3		118	112
Tm_2O_3		17	16
Yb_2O_3		105	103
Lu_2O_3		16	16
LREO	(%)	0.85	0.73
HREO		0.21	0.2
TREO		1.07	0.93

DIRECT TREATMENT OF FOXTROT ORE FOR RE RECOVERY: BATCH TESTING

The direct treatment of Foxtrot ore was investigated through a series of studies on acid baking/water leaching, solution purification, RE precipitation, RE re-dissolution and purification to remove thorium and finally RE precipitation with oxalic acid and calcination to make a mixed REO. The goal again was simplification, cost reduction and production of a premium product. The general flowsheet for the treatment scheme is shown in Figure 3.

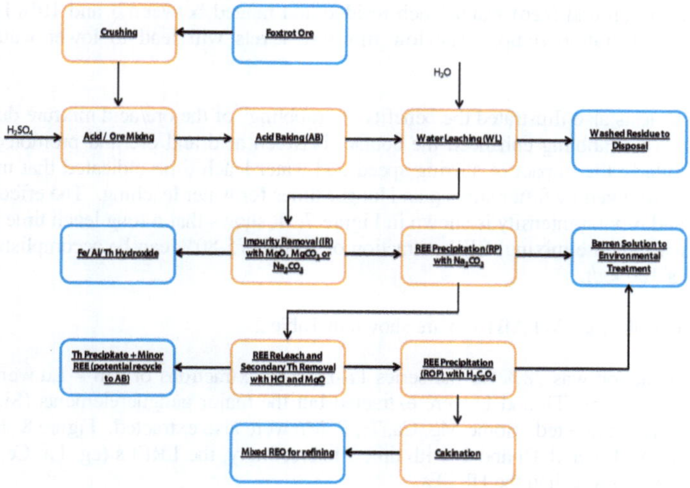

Figure 3 – Conceptual Flowsheet for Testing of the New Foxtrot Process

Acid Baking and Water Leaching

The use of acid baking and water leaching was previously tested on the Foxtrot concentrate obtained by gravity, flotation and magnetic separation. The Foxtrot RE minerals responded well to this process with high overall extractions (Dreisinger et al, 2012). The purpose of the present study was to investigate whole ore acid baking and water leaching. In addition, minimum use of acid and the impact of crush/grind size were to be investigated.

Figure 4 shows the results of the initial testing on the impact of crush size. The ore was treated with 1500 kg/t of H_2SO_4 for 4 hours at 200°C and then water leached for 24 hours to extract the REE's into solution. The acid bake extraction of REE's from the 6 mesh material was almost the same as the original concentrate (also shown on the graph). The direct extraction of the light RE's approaches 95%.

The impact of acid addition at 6 mesh is illustrated in Figure 5. At acid additions in the range of 100-250 kg H_2SO_4/t, the REE extractions were still as high as ~85% for the light REE's.

A number of experiments were done with 1 kg acid bake feed charges. The purpose was to confirm extractions and produce a larger volume of water leach solution for liquor processing. Figure 6 shows the 1 kg charge in the acid bake container before and after baking. The material is obviously coarse (6 mesh) and "dry" in appearance both before and after the acid baking and displayed a non-sticky behaviour with favourable material handling characteristics. An additional advantage is that filtration and washing are significantly reduced. For example the filtration times of the 1kg bake tests were around 5 min (at a 185 mm Buchner using W#3). Due to the coarse particle size cake, moistures are also much lower than what is normally encountered in a typical (using ground feed) water leach residue and ranged between 5 and 10% H_2O at a Buchner vacuum filtration setup. The low moisture levels will lead to lower water wash requirements.

The 1 kg "bulk" tests also illustrated the benefits of "rabbling" of the ore/acid mixture during the acid bake test. This rabbling enhanced the contact between acid and ore and promoted higher extraction. Similarly the impact of stirring speed and water leach time indicated that improved extraction was promoted by faster stirring and longer times for water leaching. The effect of WL time, density and mixing intensity is shown in Figure 7. It shows that a long leach time (24h) in combination with intense mixing, REE extraction of around 75-80% can be accomplished at an acid addition of 100 kg/t.

Detailed leach results (test WLAB16.3) are shown in Table 2.

The average extraction was 78% for the series La-Er. The extractions of Tm – Lu were lower. The radioactive elements Th and U were extracted but the major gangue elements (Si, Al, Fe, Na, K) were weakly extracted. Some Mg, Ca, Ti, P, Mn were also extracted. Figure 8 shows the extraction of the REE's and Th and U with time. Interestingly, the LREEs (eg. La, Ce, Pr, Nd) are extracted more slowly than the HREEs.

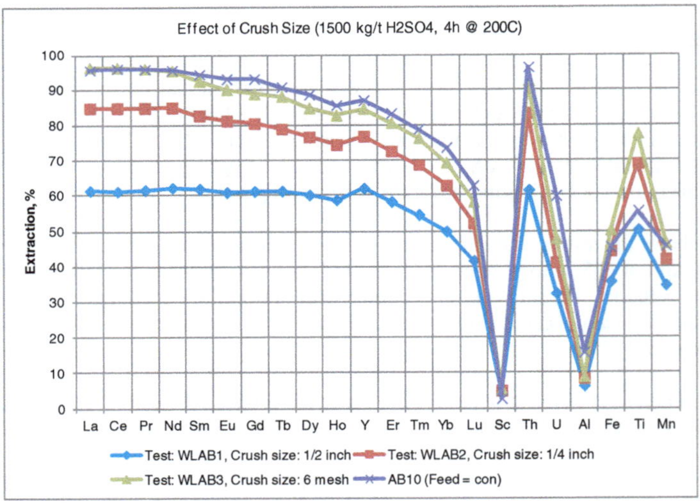

Figure 4 –Acid Bake-Water Leach REE Extraction - Effect of Crush Sizes

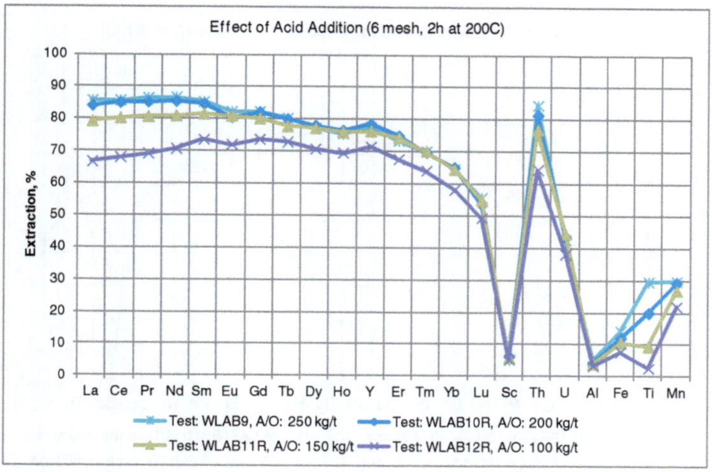

Figure 5 – Acid Bake-Water Leach REE Extraction - Effect of Acid Addition

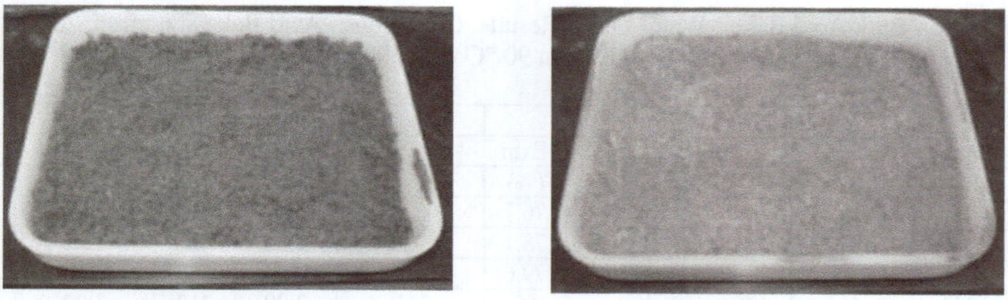

BEFORE ACID BAKE

AFTER ACID BAKE

Figure 6 – Photograph of acid bake feed and product. Conditions: 6 mesh, 100 kg/t H_2SO_4, 200 °C, 2 hrs.

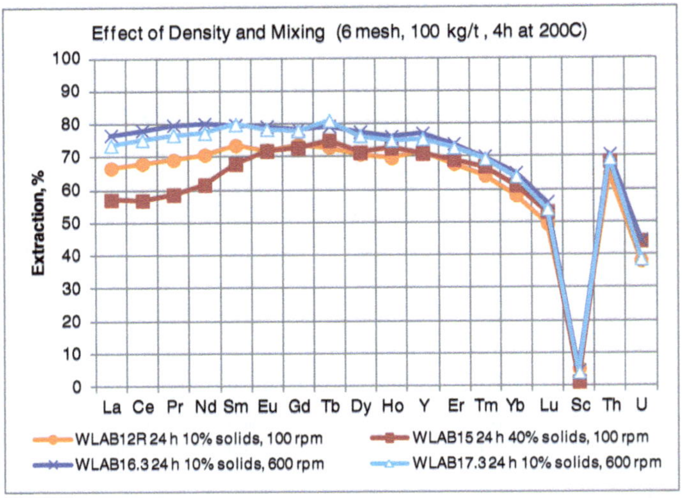

Figure 7 – Acid Bake-Water Leach REE Extraction - Effect of Pulp Density and Mixing

Table 2 – Bulk Acid Bake - Water Leach Results. Conditions: Acid Bake at 6 mesh, 200 °C, 2 hours and 24 h Water Leach at 90 °C with 600 rpm mixing intensity.

Rare Earth Extraction					Impurity Extraction				
Element	Analysis (mg/L or g/t)			Extn	Element	Analysis (mg/L, % or g/t)			Extn
	Feed	PLS	Residue	(%)		Feed	PLS	Residue	(%)
La	1720	144	392	76.7	Th	109	9.35	34.6	70.4
Ce	3720	321	805	78.1	U	22.4	1.2	13.5	44.4
Pr	437	39.7	90.8	79.6	Si	31.32	288	32.44	0.8
Nd	1610	148	330	80	Al	3.99	212	3.97	4.6
Sm	297	27.9	63.8	79.6	Fe	7.83	496	7.69	5.5
Eu	15.5	1.51	3.6	79.1	Mg	0.12	43.5	0.07	36.7
Gd	244	22.6	56.1	78.4	Ca	1.45	642	0.96	40
Tb	37.3	3.59	8.5	79.2	Na	2.13	47	2.14	2.6
Dy	223	20.8	54.6	77.5	K	3.36	384	3.44	9.2
Ho	43.7	4.09	11.7	76	Ti	0.27	4.6	0.28	1.5
Y	1090	107	288	77	P	0.01	5	0.02	33.5
Er	122	11.2	36	73.8	Mn	0.23	82.3	0.19	27.6
Tm	17.2	1.49	5.8	69.9					
Yb	111	8.56	41.8	64.9					
Lu	15.8	1.02	7.3	55.9					

Impurity Removal by Oxidation and Precipitation

The initial removal of impurities was tested by pH adjustment and oxidation (for Fe removal). Three alkalis were tested including MgO, $MgCO_3$ and Na_2CO_3. For each test, the removal of impurities appeared to be maximized with minimum rare earth loss due to co-precipitation. All three alkalis were successful. Magnesium carbonate ($MgCO_3$) was selected for a bulk impurity removal test. A volume of ~13 L of water leach solution was prepared and heated to 75 °C and treated with ~ 0.5 g/L of H_2O_2 to raise the ORP to +600 mV (vs Ag/AgCl). The pH was then adjusted to 3.75 with 15 % solid slurry of $MgCO_3$ and held for 1 h. The impurity precipitates were filtered and washed. The results are summarized in Table 3 below. More than 90% of the iron was eliminated along with 88.4% of the thorium. There was also significant rejection of Si, Al, Ti and P. The losses of REEs ranged from 0.74 to 3.6% from La to Lu. Note that the final precipitate was analyzed at 0.018% Mg, indicating a high efficiency of $MgCO_3$ use.

Bulk Rare Earth Precipitation

The purified solution was treated with a soda ash solution (Na_2CO_3) to precipitate the REEs into a mixed carbonate product for further purification. This was a further change in procedure from the earlier study (Dreisinger et al, 2012) where oxalic acid was added as the primary REE precipitant. A pH target of 7.25 at ambient temperature was set. The results are shown in Table 4 below. The precipitation of REEs approaches 100%. The co-precipitation of Th, U, Fe, Al is similarly very high. The mixed REE carbonate precipitate may be further refined by a re-leach, oxalate precipitation and calcination method to form a mixed REO for refining. The overall recovery of REEs from ore to mixed carbonate precipitate has been calculated and summarized in Table 5.

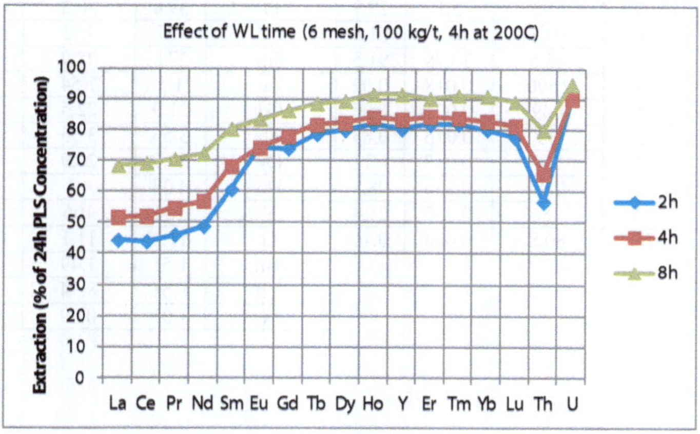

Figure 8 – Acid Bake-Water Leach REE Extraction - Effect of Time

The final weight of the calcine was less than 10 g. This made it difficult to accurately analyse the final product. Both ICP and XRF analytical techniques were used and gave some variation in the individual rare earth values. Further work is underway on a larger scale test of the Search Minerals Whole Ore Rare Earth Recovery Process, which will provide larger amounts of calcine for analysis. It is still however possible to make some general comments about the assays.

1. The calcine is generally very low in contamination and mostly consists of rare earth oxides.
2. The thorium content of the precipitate was 3.6 g/t and the calcine was reported at 6.5 g/t Th by ICP, which is a low value for a rare earth oxide.
3. The uranium content of the calcine is <50 g/t.
4. The aluminum and iron values are suitably low at 146 and 9 g/t in the precipitate. The iron in the calcine was analyzed by ICP at 744 g/t. There is no explanation as to why this increased.
5. Minor amounts of alkali, alkaline earth and base metals are present. This is probably due to some over addition of oxalic acid during precipitation. I.e. the rare earths are precipitated first by the oxalic acid followed by other metals (Mg, Ca, K, Mn, Zn).
6. The F content at 0.15% is higher than expected. This may be minimized by more selective oxalate precipitation.
7. The total carbon content was very low at 0.02% indicating that the calcination was complete.

Table 3 – Impurity Removal Results. Conditions: pH 3.75 with $MgCO_3$ addition for 1 h at 75 °C

Impurity Rejection					Rare Earth Co-Precipitation				
Element	Analysis (mg/L or g/t)			Ppt	Element	Analysis (mg/L or g/t)			Ppt
	PLS	Filtrate	Prec.	(%)		PLS	Filtrate	Prec.	(%)
Th	9.43	1.07	4810	88.4	La	141	145	643	0.74
U	1.21	1.15	36.5	5.09	Ce	314	324	1960	1.01
Si	279	202	5.1	29.9	Pr	39.4	40.2	286	1.19
Al	201	99.1	6.35	51.8	Nd	147	152	1120	1.23
Fe	619	48.5	30.78	91.5	Sm	27.8	28.9	254	1.46
Mg	35.7	1690	0.018	0.02	Eu	1.5	1.54	13.2	1.43
Ca	602	649	0.021	0.06	Gd	21.9	22.5	210	1.55
Na	28	30	0.015	0.83	Tb	3.55	3.55	37.6	1.76
K	238	259	0.017	0.11	Dy	20.7	20.9	224	1.78
Ti	5.68	0.07	0.336	98.8	Ho	4.08	4.1	45.4	1.84
P	5	5	0.083	21.4	Y	104	106	875	1.38
Mn	76.2	80.8	0.008	0.16	Er	11.1	11.1	156	2.32
					Tm	1.51	1.49	26.7	2.94
					Yb	8.51	8.34	185	3.61
					Lu	1.01	0.99	21.8	3.58

Table 4 – REE Precipitation Results. Conditions: pH 7.25 with Na_2CO_3 addition for 3 h at 25 °C

Rare Earth Precipitation					Impurity Precipitation				
Element	Analysis (mg/L or g/t)			Pptn	Element	Analysis (mg/L, % or g/t)			Pptn
	Feed Soln	Filtrate	Ppte	(%)		Feed Soln	Filtrate	Ppte	(%)
La	134	0.24	57700	99.8	Th	0.84	0.04	422	95.7
Ce	315	0.39	129000	99.9	U	1.07	0.13	467	88.4
Pr	36.3	0.06	15600	99.8	Al	83.6	0.05	3.67	99.4
Nd	136	0.18	58800	99.9	Fe	39.3	0.2	1.81	99.5
Sm	26.1	0.05	11000	99.8	Mg	1550	1560	0.505	0.7
Eu	1.39	0.03	584	97.6	Ca	616	583	1.95	6.6
Gd	23	0.04	10900	99.8	Na	191	1050	0.125	0.3
Tb	3.62	0.03	1640	99.1	K	298	289	0.03	2.2
Dy	20.8	0.05	9270	99.7	P	5	5	0.003	1.2
Ho	4.08	0.02	1800	99.5	Mn	77.4	68.8	0.131	3.9
Y	107	0.3	42100	99.7					
Er	11.1	0.04	4880	99.6					
Tm	1.46	0.04	636	97.1					
Yb	7.92	0.03	3470	99.6					
Lu	0.93	0.03	409	96.6					

Table 5 – Overall Recovery (%) of Rare Earth Elements from Foxtrot Ore to Mixed Carbonate Precipitate

La	Ce	Pr	Nd	Sm	Eu	Gd	Tb	Dy	Ho	Y	Er	Tm	Yb	Lu
75.9	77.3	78.9	79.3	78.9	78.4	77.6	78.4	76.7	75.3	76.3	73.1	69.2	64.3	55.4

Mixed Carbonate Re-leach, Thorium Removal, REE Oxalate Precipitation and Calcination

The mixed carbonate product was re-leached with 37% HCl solution at pH 1 for 1hour at 80 °C. The chloride leach solution was then treated with 15 % slurry of MgO in deionized water to pH 3.8 at 50 °C for 1 h to reprecipitate re-leached thorium. The re-leach process was virtually 100% effective in re-dissolving the REEs. The re-leach residue and the thorium removal residue would be returned to the water leach process to minimize any REE loss. The low-thorium solution was then treated with oxalic acid to selectively precipitate the REEs. The rare earth oxalate was then calcined at 750 °C for 4 h to produce a final mixed rare earth product. The assay of the rare earth oxalate and the calcined rare earth oxalate are shown in Table 6.

DIRECT TREATMENT OF FOXTROT ORE FOR RE RECOVERY: CONTINUOUS KILN TESTING

A larger scale test of the Search Process was conducted by crushing a sample of Foxtrot to 6 mesh and premixing with acid in a cement mixer configuration (100 kg/t of H_2SO_4). The acid-ore mixture was then fed at 2 kg/h by a screw conveyor to a continuous kiln system operated at 190.7 °C bed temperature. The residence time was 1.4 h in the hot zone of the kiln (3 h total). The calcine was water leached at 90 °C for 24 at 10% solids. The leachate was then processed via impurity removal, carbonate precipitation, hydrochloric acid re-leaching and purification and oxalate precipitation and calcination. The product of this treatment was analyzed and is reported in Table 7.

The final product from the continuous kiln process (full flowsheet) was a mixed rare earth oxide product containing 98.9% total rare earth oxide (TREO). The thorium content of the product was very low at 2.1 g/t with generally low content of all key impurities.

Table 6 – Chemical Analysis of the Rare Earth Oxalate and Rare Earth Calcine Products from Small Scale Batch Testing

Rare Earth				Impurity			
Element	Analysis (g/t)			Element	Analysis (% or g/t)		
	Oxalate-ICP	Calcine-ICP	Calcine XRF		Oxalate-ICP	Calcine-ICP	Calcine XRF
La	134	0.24	57700	Th	0.84	0.04	422
Ce	315	0.39	129000	U	1.07	0.13	467
Pr	36.3	0.06	15600	Al	83.6	0.05	3.67
Nd	136	0.18	58800	Fe	39.3	0.2	1.81
Sm	26.1	0.05	11000	Mg	1550	1560	0.505
Eu	1.39	0.03	584	Ca	616	583	1.95
Gd	23	0.04	10900	Na	191	1050	0.125
Tb	3.62	0.03	1640	K	298	289	0.03
Dy	20.8	0.05	9270	P	5	5	0.003
Ho	4.08	0.02	1800	Mn	77.4	68.8	0.131
Y	107	0.3	42100	Zn	804	1490	
Er	11.1	0.04	4880	F		0.15	
Tm	1.46	0.04	636	Ctot		0.02	
Yb	7.92	0.03	3470				
Lu	0.93	0.03	409				

Table 7 – Chemical Analysis of the Mixed REO Calcine Product from the Kiln Testing

Rare Earth		Impurity	
Element	Analysis (%)	Element	Analysis (g/t)
La	12.0	Th	2.1
Ce	31.4	U	96.6
Pr	3.87	Si	93.5
Nd	15.4	Al	<53
Sm	2.53	Fe	69.9
Eu	0.147	Mg	92
Gd	2.13	Ca	1572
Tb	0.314	Na	<300
Dy	2.00	K	<83
Ho	0.413	Ti	<60
Y	9.8	P	43.6
Er	1.07	Mn	465
Tm	0.156	Zn	<200
Yb	0.705	F	290
Lu	0.0887	Ctot	<100
TREE	82.0	S	<100
LREE	65.2		
HREE	16.8		
TREO	98.9		

CONCLUSIONS

Search Minerals has rapidly advanced the exploration and development of the Port Hope – Simpson Rare Earth Element District. The Foxtrot orebody contains a variety of rare earth minerals and is rich in heavy rare earth elements. A simple direct leaching process has been developed for treatment of Foxtrot ore that represents a breakthrough in the development of this orebody. The process involves crushing to 6 mesh particle size, application of 100 kg/t of H_2SO_4 to the ore at 200 °C for 2 hours followed by a 24 h water leach to produce a weakly acidic product leach solution. Favourable materials handling characteristics of the acid/feed mixture and calcine as well as easy filtration characteristics of the water leach have led to the development of a much more operable acid bake circuit than other "typical" REE acid bake circuits.

After oxidation and pH adjustment with $MgCO_3$ slurry to precipitate the bulk of the iron and thorium from the solution, sodium carbonate was added to precipitate a mixed carbonate rare earth product. The mixed carbonate was then re-leached with hydrochloric acid at pH 1 to produce a strong rare earth chloride solution. The pH was again increased with MgO to reject residual amounts of thorium and other impurities. The rare earths were re-precipitated with oxalic acid addition to produce a high quality mixed rare earth oxalate for calcination. The oxalate precipitate was calcined at 750 °C to produce a high quality mixed rare earth oxide product suitable for refining. This product could be re-leached at a refinery with HCl to produce a high strength solution for rare earth separation by multi-stage solvent extraction processing.

The final mixed rare earth oxide product from a continuous kiln treatment of an acid ore mixture, followed by the purification steps analyzed 98.9% TREO with low impurities. Search Minerals is continuing to invest in mineral exploration in the Port Hope Simpson district. A 40 t sample of Foxtrot mineralization has been extracted for a large scale pilot plant study expected in Q4-2015 through Q2-2016. Engineering and cost studies are continuing.

ACKNOWLEDGEMENTS

Support was received from the Research & Development Corporation ("**RDC**") of Newfoundland and Labrador and from the Atlantic Canada Opportunities Agency ("**ACOA**") for this development.

REFERENCES

Srivastava, R.M., Gauthier, J., Cox, J., Krutzelmann, H. (2013), *Technical Report on the Foxtrot Project in Labrador, Newfoundland and Labrador Canada* (Report NI 43-101), Toronto, ON: Roscoe Postle Associates Inc., Retrieved from http://www.sedar.com

Srivastava, R.M., & Gauthier, J. (2012). *Search Minerals Inc. Technical Report on the Foxtrot Project in Labrador, Newfoundland and Labrador Canada.* (Report NI 43-101). Toronto, ON: Roscoe Postle Associates Inc. Retrieved from http://www.sedar.com

Dreisinger, D.B., Clucas, J.D., Verbaan, N., Grammatikopoulos, T., Aghamirian, M., Forstner, C., "The Processing of REE's from Search Minerals Foxtrot Resource", Proceedings of Rare Earths 2012, Eds, J.R. Goode, G. Moldoveanu, M.S. Reyat, CIM Metsoc (Montreal), 81-94.

Dreisinger, D.B., Verbaan, N., Johnson, M., "The Processing of REE's from Search Minerals Foxtrot Resource – an Update", COM 2014 - Conference of Metallurgists Proceedings (2014), 13 pages.

Rare Metal Technology 2016
Edited by: Shafiq Alam, Hojong Kim, Neale R. Neelameggham, Takanari Ouchi, and Harald Oosterhof
TMS (The Minerals, Metals & Materials Society), 2016

HYDROMETALLURGICAL EXTRACTION OF RARE EARTH ELEMENTS FROM LOW GRADE MINE TAILINGS

S. Peelman[1], Z.H.I. Sun[1], J. Sietsma[1], Y. Yang[1]

[1]Department of Materials Science and Engineering, Delft University of Technology;
Mekelweg 2, 2628 CD Delft, The Netherlands

Keywords: Rare earths, Apatite, Phosphoric acid production, Low grade REE resources

Abstract

The rare earth elements (REEs) present in the Kiruna iron ore mine tailings are being considered as a new REE resource for the EU. These tailings, after beneficiation, contain approximately 5000 ppm of rare earths, concentrated in apatite and monazite minerals. To economically extract the REEs from these tailings the phosphorous contained within the apatite must also be extracted and the waste production of the process should be minimal. To achieve the extraction of REEs and phosphorous HCl and HNO_3 were investigated as possible leaching agents. Based on the results a leaching process using HNO_3 is proposed. This process produces H_3PO_4 from the apatite while simultaneously dissolving the REEs. The resulting leach liquor is then cooled to remove the unwanted Ca as $Ca(NO_3)_2$. Analysis of the leach liquor reveals extraction ratios of between 75% and 100% for the heavy REE (e.g. Y, Eu, Dy, …) and below 40% from the light REEs (i.e. Ce, La). Full phosphorous extraction was also achieved.

Introduction

The rare earth elements (REEs) are a group of 17 chemically similar elements comprising of the lanthanides, Y and Sc. These elements have become increasingly important in the modern society, with a variety of applications in electronics (e.g. FeNdB magnets), chemical industry (e.g. REE catalysts) and even the energy industry (e.g. batteries and windmills) [1]-[2]. Due to the importance of these elements they have been classified as critical resources within the EU [3] and a shortage of these elements is expected. The expected shortage is not due to the scarcity of these elements, because unlike the name suggest REEs are not rare, they are in fact quite common [4]. The difficulty lies in the fact that REEs have a tendency to substitute with common rock forming elements. This causes the REEs to have only a limited amount of ores where their concentration is high enough for economical extraction [5]. The major mines where these ores are being exploited are mostly located in China, in fact China is the largest producer of REEs globally with 95-97% of production located there [6]. The EU wishes to secure a source of REEs prevent an uncertain supply.

The EU currently has no active mines that produce REEs. To secure an REE supply the EU has been exploring readily available REE sources of lower grade, which can be processed quickly without the need to excavate a completely new mine. One such potential source was found in the mine tailings from the LKAB (Luossavaara-Kiirunavaara Aktiebolag) iron ore mine in Kiruna, Sweden. These tailings have been found to contain a substantial amount of apatite, a mineral that is known to be associated with REEs [5]. Analysis of these tailings revealed the presence 1200-

1500 ppm REEs, which considering the 50-100M tons of tailings in the pond, represents the largest secondary REE deposit in Europe. In the recently EU funded project REEcover [7], the evaluation of the feasibility of extraction of REEs from these tailings is being conducted.

The recovery of REEs from these low grade tailings faces not only a technological challenge, but an economical one as well. Due to the instability of the REE market it is difficult to secure a profitable secondary production of the REEs, to the point that even primary mines are not operating profitably, such as the Mountain Pass mine [8]. Therefore all valuable elements contained within the tailings must be extracted in order for this to be an even remotely viable REE resource. As mentioned the tailings contain a high amount of apatite, which is one of the primary resources of phosphorous in the world. Phosphorous is also regarded as a critical element by EU and is there for just as valuable as the REEs. Hence these tailings should be considered as a potential resource for REEs and P and any extraction process applied to them should extract both elements. This is what is being researched in this work: a process for the extraction of REEs and P from the Kiruna mine tailings.

The primary way to valorize P from apatite is to produce H_3PO_4. Traditionally this is done by dissolving the apatite with H_2SO_4 to form H_3PO_4 as shown in reaction (1).

$$Ca_5(PO_4)_3F + 5\,H_2SO_4 \rightarrow 5\,CaSO_4 + 3\,H_3PO_4 + HF \qquad (1)$$

This process has two major problems: it produces copious amounts of waste, 5 tons per ton of produced P_2O_5, in the form of gypsum ($CaSO_4$) and 80% of the dissolved REEs are also lost to the gypsum [9]. These two problems eliminate H_2SO_4 as a possible leachant for the extraction of P and REEs. In the present work the other two common mineral acids, HCl (reaction 2) and HNO_3 (reaction 3), are tested to observe their potential as leaching agents. HNO_3 offers the potential to also valorize the Ca of the apatite as $Ca(NO_3)_2$ adding an additional value stream to the process [10]. The downside of HNO_3 is that this will also produce nitrated waste water. HCl does not produce this type of waste water, but also does not allow for Ca valorization, as $CaCl_2$ cannot be easily precipitated.

$$Ca_5(PO_4)_3F + 10\,HCl \rightarrow 5\,CaCl_2 + 3\,H_3PO_4 + HF \qquad (2)$$

$$Ca_5(PO_4)_3F + 10\,HNO_3 \rightarrow 5\,Ca(NO_3)_2 + 3\,H_3PO_4 + HF \qquad (3)$$

For hydrometallurgical processing such tailings, a detailed understanding of the process chemistry and engineering is required. This research provides investigations about the effects of acid type and acid concentrations on the hydrometallurgical leaching the mine tailings. The leaching behaviors of both REEs and P are preliminarily understood, with the aim to develop a process for critical elements recovery from this kind of secondary resources.

Experimental

Material

The input material for this work was provided by partners of the REEcover project, (led by prof. Pålsson) of Luleå University of Technology (LTU). It came from a LKAB pilot campaign to produce an apatite concentrate. The material was delivered as a powder with particle size ranging from approximately 10 to 100 μm.

Leaching

Leaching was performed in a 1 liter glass reactor equipped with a glass paddle stirrer and submerged in a water bath for heating. The stirring rate was kept constant for all experiments at 600 rpm. A condenser was used for high temperature experiments to prevent solvent loss. Leach residues were filtered using a Buchner filter with Whatman 52 filtration paper (7μm) and washed with demineralized water. Filtered leach residues were dried in a drying oven at 105°C for 24h.

All experiments were performed under normal atmosphere and using conventional equipment. High temperature (above 100°C) and high pressure leaching in autoclaves was not considered as the value contained within these tailings is too low to justify the use of such equipment.

The chemicals used in the experiments were: 37wt.% HCl (ACS grade) and 65wt.% HNO_3 (ACS grade) from Sigma-Aldrich, $Ca(NO_3)_2 \cdot 4H_2O$ (ACS grade 99-103% purity) from Alfa Aesar and demineralized water.

Characterization of the Input Material

Chemical Analysis. Chemical analysis of the input material was performed using Inductive Coupled Plasma – Optical Emission Spectroscopy (ICP-OES) (Spectro Arcos EOP), Inductive Coupled Plasma – Mass Spectroscopy (ICP-MS) (ThermoFischer Scientific X-series quadrupole ICP-MS) and X-ray Fluorescence (XRF) (Panalytical Axios Max WD-XRF). To ensure total dissolution of the entire sample for ICP-OES and -MS analysis, digestion with alkaline flux was chosen. 1.5g of the input material was digested using 15g of borate flux of 50:50 $Na_2B_4O_7$ and Na_2CO_3 at 1000°C for 1h. The resulting glass was subsequently dissolved in diluted HNO_3 and diluted further for analysis. The solution was measured with ICP-OES and ICP-MS to identify the main (OES) and trace elements (MS). Due to lack of standards, Si and F were measured with XRF on the solid material, not ICP-OES. The results of the chemical analysis can be seen in Tables I and II. Only elements with a concentration above 3000 ppm are considered as main elements.

Table I: Main elements of the input material

Main element concentration (%) of the input material						
Ca	**P**	**Fe**	**Mg**	**Al**	**Si***	**F***
37.6	15.7	1.15	0.84	0.29	0.59	1.99

*measured with XRF

Table II: REE concentrations in the input materials

REE concentration (ppm) of the input material							
La	**Ce**	**Pr**	**Nd**	**Sm**	**Eu**	**Gd**	**Tb**
902.9	2047.1	238.5	906.0	147.5	19.1	137.7	16.2
Y	**Dy**	**Ho**	**Er**	**Tm**	**Yb**	**Lu**	**SUM**
664.1	100.7	17.3	54.1	5.3	38.3	4.0	5298.7

The input material contains approximately 5300 ppm of REEs, of which 55% are Ce and La, 22% are Nd and Pr and 23% are heavy REEs (mainly Y at 13%).

Mineral Identification. X-ray diffraction (XRD) (Bruker D8 Advance diffractometer Bragg-Brentano geometry and Lynxeye position sensitive detector) analysis was used to identify the main minerals present in the input material. Based on the spectrum, Figure 1, the main minerals were identified as (fluor)apatite ($Ca_5(PO_4)_3F$), calcite ($CaCO_3$) and dolomite ($CaMg(CO_3)_2$). Minor signals for quartz and magnetite were identified as well.

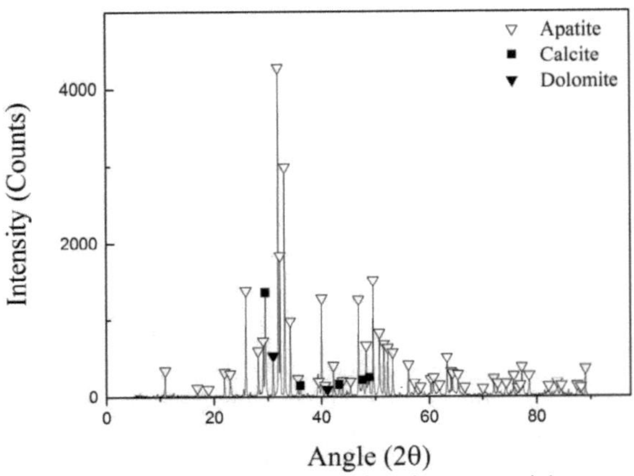

Figure 1: XRD spectrum of the input material

Based on the chemical composition of the material, the distribution of the main minerals is approximately 83wt.% apatite, 6wt.% calcite and 6wt.% dolomite. The remaining 5% consists of mainly quartz and magnetite together with minor fractions of other refractory minerals. From this analysis we conclude that the input material is an apatite concentrate and will be referred to as such in the remainder of this work. No REE minerals could be identified with the XRD measurement of the apatite concentrate. This signifies that the REE mineral concentration in the apatite concentrate is below the detection limit of the XRD (≈1%).

REE Mineral Identification. To confirm the presence of REE minerals in the apatite concentrate the material was analyzed with Scanning Electron microscopy (SEM) (JEOL JSM 6500F). Using Back Scattered Electron (BSE) imaging it is possible to trace particles which are composed of heavy elements. Heavier elements are more likely to cause backscattering of electrons, making them appear brighter on the detector [11]. Lanthanides have atomic numbers ranging from 57 (La) to 71 (Lu), this makes particles that contain them appear brighter than the Ca (atomic number 20) containing particles. Using this principle it is possible to identify individual REE particles within the concentrate and analyze their composition with Energy-dispersive X-ray spectroscopy (EDS), see Figure 2a.

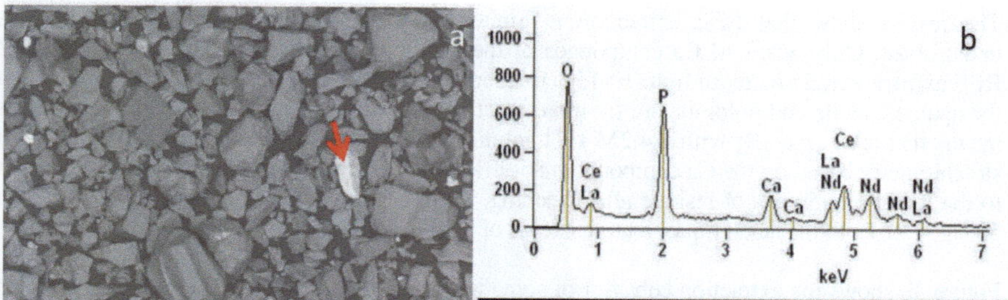

wt.% measured with EDS					
O-K	P-K	Ca-K	La-L	Ce-L	Nd-L
28.9	13.9	5.04	11.17	30.61	10.4

Figure 2: a: SEM image of input material using BSE imaging; b: EDS analysis of indicated particle (red arrow)

EDS analysis of the indicated particle identifies it as a REE mineral (see Figure 2b) and its chemical composition suggests that this mineral is a monazite mineral ((Ce,La,Nd,...)PO_4), which agrees with the mineralogical study performed by Pålsson et al. [12]. Further analysis of the other bright particles shows identical EDS spectra.

Results And Discussion

Leaching of the Apatite Concentrate

The apatite material was leached with 250 ml HCl solutions of increasing concentration. The concentrations used were 0.3M, 1.5M, 2.1M, 4.2M and 6.3M HCl. A liquid solid ratio (L/S) of 5 was maintained for these experiments. The experiments were run for 6h at room temperature. The results of these experiments are displayed on Figure 3a and Figure 3b.

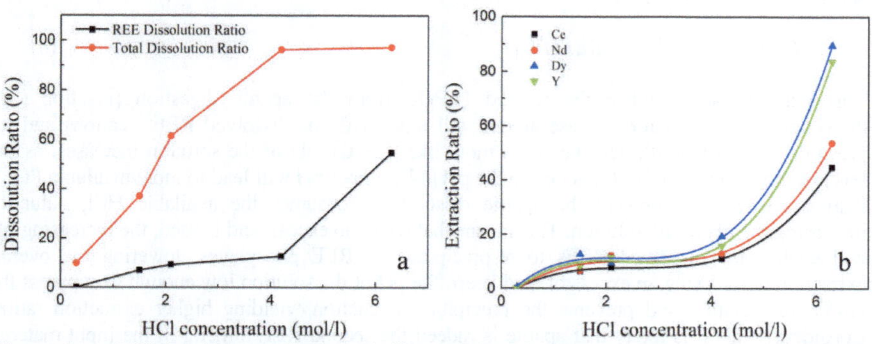

Figure 3: a: HCl leaching profile - total dissolution and REE dissolution; b: individual REE extraction ratios as a function of HCl concentration

The results show that REE extraction is limited as long as there is apatite that remains undissolved. Only when all Ca compounds of the apatite concentrate have dissolved, does the REE extraction ratio increase from 10-15% to about 55% (see Figure 3a). This clearly shows that the apatite, calcite and dolomite are far more reactive than the REE minerals, i.e. monazite. This means that after leaching with a 4.2M HCl solution (10% excess over the required amount for stoichiometry based on the Ca compounds) 85-90% of the REEs in the apatite concentrate report to the remaining 5wt.% of residue after leaching. To extract any significant amount (more than 50%) of REEs to the leach liquor a large excess of acid is required.

Figure 3b shows the extraction behavior of some individual REEs. A clear increase in extraction ratio is observed when comparing the heavy REEs (here Y at 84% and Dy at 90%) to the lighter REEs (Ce at 45% and Nd at 54%) when leaching with a large excess of acid. This behavior is also observed at the lower acid concentrations, albeit in far smaller amounts (Ce at 11% and Dy at 20%). This could be attributed to the marginally higher solubility of the heavy REEs [13] or could indicate that the heavy REE and light REE are associated with different minerals within the apatite concentrate. Based on the EDS analysis of the apatite concentrate (Figure 2b) one of these minerals is most likely monazite. Previous studies have shown that monazite is light REE mineral and that in primary monazite processing it requires very aggressive reaction conditions to dissolve the monazite [8]. A possibility of the other REE mineral is the apatite itself. Previous studies have shown that the apatite mineral can also contain REEs [14]. Through substitution of Ca REEs can be incorporated in the apatite lattice. Thus as the apatite dissolves the REEs in its lattice will co-dissolve.

However at first glance, this behavior is not observed in the HCl leaching profile (Figure 3a and 3b). The REE extraction ratio remains virtually constant as the apatite dissolution ratio increases up until full apatite dissolution (4.2M) is achieved. The REE extraction ratio only increases when all of the apatite has been dissolved and excess acid is present (6.3M). However, there is a side reaction that should be accounted for, REE phosphate precipitation (reaction (4)). REE phosphates are very insoluble, as shown by the equilibrium constant of reaction (4), approximately $K = 10^{22}$ [15].

$$REE^{3+} + PO_4{}^{3-} \rightarrow REE(PO_4) \tag{4}$$

This reaction occurs when the formed H_3PO_4, from the apatite digestion (reaction (2)), dissociates to PO_4^{3-} anions. These anions will react with the dissolved REE^{3+} cations and re-precipitate as $REE(PO_4)$. This becomes more likely as the pH of the solution increases, as the level of dissociation of H_3PO_4 is tied to the pH [14], higher pH will lead to more available PO_4^{3-}. During the leaching process the apatite dissolution consumes the available HCl, naturally increasing the pH of the solution. This means that when no excess acid is used, the increasing pH will enable any dissolved REEs to re-precipitate as REE phosphates, lowering the overall extraction ratio. Using an excess of acid keeps the pH of the solution low enough to suppress the H_3PO_4 dissociation and prevents the precipitation reaction, yielding higher extraction ratios. Considering this it is likely that apatite is indeed the second REE mineral in the input material and that it's the source of the heavy REEs.

Increasing the Acid Concentration with HNO_3

Based on the leaching results of the HCl leaching, a large excess of acid is required to achieve a moderate extraction ratio for the REEs. Based on this observation the leaching experiment was repeated with concentrated HNO_3 (65wt.%) instead of HCl (37wt.%) to achieve an as high as possible acid concentration. To further increase the reactivity the reaction temperature was raised to 60-70°C. The choice to use HNO_3 is not only based on the increased acid concentration, HNO_3 also has the added benefit of enabling $Ca(NO_3)_2$ precipitation after leaching, as demonstrated by the Norsk Hydro process [16]. This allows for valorization of the Ca as $Ca(NO_3)_2$, which is a resource for the fertilizer industry [10]. This improves the economy of the process and also benefits the recovery of REEs from the solution at a later stage, as most of the Ca will have been removed already. Taking the precipitation of $Ca(NO_3)_2$ into account the L/S of the experiment was lowered to 2 to facilitate the Ca removal afterwards.

Table III: REE extraction ratios HNO_3 leaching (L/S=2 and T=60-70°C)

Extraction ratio REE HNO_3 leaching L/S=2 (%)							
La	**Ce**	**Pr**	**Nd**	**Sm**	**Eu**	**Gd**	**Tb**
37.8	26.0	45.2	46.9	59.3	66.5	68.3	73.2
Y	**Dy**	**Ho**	**Er**	**Tm**	**Yb**	**Lu**	
73.6	79.1	89.0	75.9	100.0	80.6	100.0	

Table III shows the REE extraction ratios of the HNO_3 leach at elevated temperatures. These extraction ratios show that the increased acid concentration does not increase the REE extraction ratio, nor does the elevated temperature. In fact the extraction ratios are 5% lower than what they were with the HCl leaching. This could be attributed to the lower L/S and the decreased REE solubility at higher temperature [8], which can facilitate the re-precipitation of the REEs.

The lack of increase in REE extraction indicates that the increased acid concentration does not dissolve any additional REEs. This means that the remaining REEs are locked in a phase that does not react with HNO_3 or the HCl under the present conditions. This indicates that there are indeed 2 different REE phases in the apatite concentrate: one which is rich in heavy REEs (apatite) and is dissolved by the acid and the other is rich in light REEs (monazite) and is not dissolved by the acid.

Leaching Residue Analysis

XRD Analysis of the Leach Residues. The residues of the leaching experiments were analyzed with XRD, and SEM/EDS to identify the minerals resistant to dissolution. XRD analysis of the 4.2M HCl leach residue (Figure 4) confirms that all apatite, calcite and dolomite have dissolved. It has also identified the major refractory minerals that were not detected on the XRD analysis of the material prior to leaching. Most of these refractory minerals are magnesio-silicates and iron oxides.

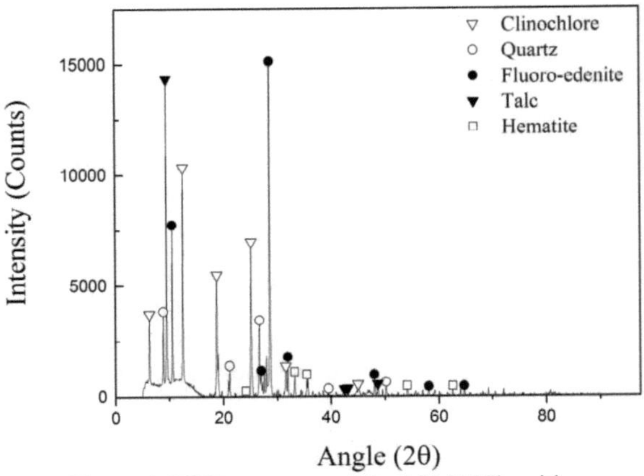

Figure 4: XRD spectrum of the 4.2M HCl residue

XRD analysis of the HNO_3 residue (Figure 5) identifies additional refractory minerals, as well as detecting monazite as part of the residue. This proves that the observed particles with SEM are indeed monazite and that monazite is resistant to acid digestion under atmospheric conditions.

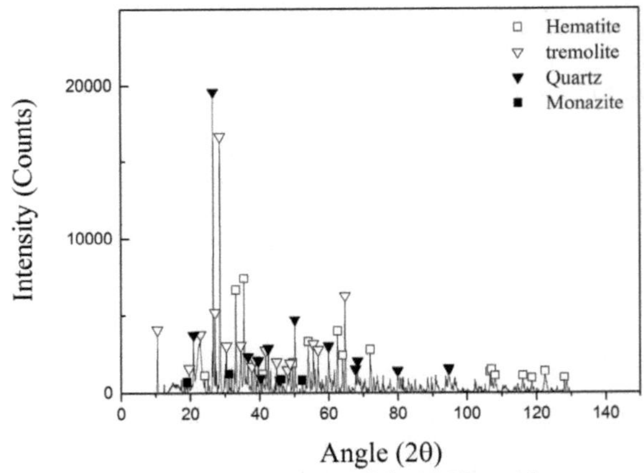

Figure 5: XRD spectrum of the HNO_3 residue

SEM/EDS Analysis Of The Leach Residues. SEM/EDS analysis of the 6.3M HCl (Figure 6) and the HNO_3 (Figure 7) leaching residues reveal the presence of a substantial amount of monazite.

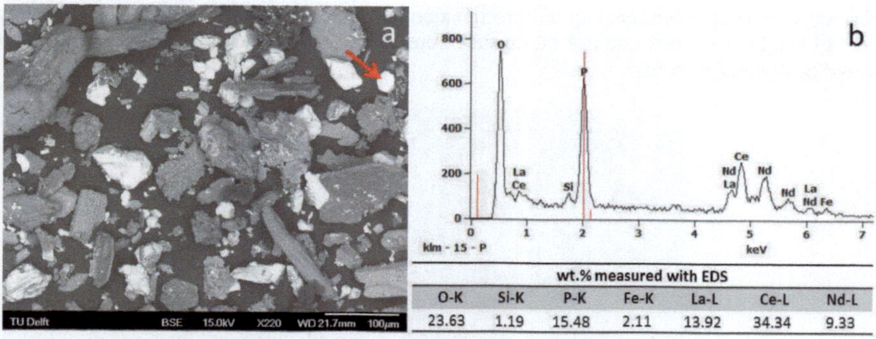

wt.% measured with EDS						
O-K	Si-K	P-K	Fe-K	La-L	Ce-L	Nd-L
23.63	1.19	15.48	2.11	13.92	34.34	9.33

Figure 6: a: SEM image of the 6.3M HCl residue using BSE imaging; b: EDS analysis of indicated particle (red arrow)

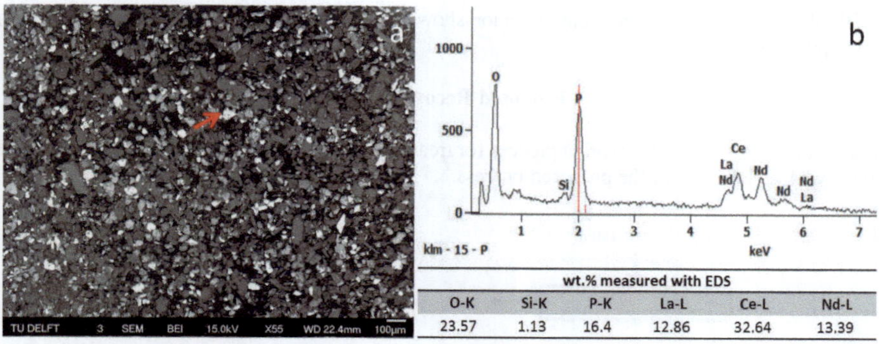

wt.% measured with EDS					
O-K	Si-K	P-K	La-L	Ce-L	Nd-L
23.57	1.13	16.4	12.86	32.64	13.39

Figure 7: a: SEM image of the HNO_3 residue using BSE imaging; b: EDS analysis of indicated particle (red arrow)

The monazite particles in both residues appear smooth and shown no sign of partial dissolution. The observable particles have an approximate particle size between 10-50μm, the same size as the observed particles in the material prior to the leach (Figure 2a). This proves that the monazite is not dissolving under these leaching conditions. Given that monazite is known to be a REE mineral that mainly contains light REEs [8], it's possible to conclude that the heavy REEs that are extracted during the 6.3M HCl and HNO_3 leaching experiments are not associated with the monazite.

<u>Precipitation Of $Ca(NO_3)_2$</u>

Based on an existing H_3PO_4 production process by Norsk Hydro, that uses HNO_3 instead of H_2SO_4 [16], $Ca(NO_3)_2$ can be precipitated from the leach liquor by cooling the liquor down to minus 5°C. To confirm this behavior and to observe if it affects the REEs in solution 20ml samples of the HNO_3 leach liquors were cooled, by means of a freezer, to minus 20°C for 24h. Initial tests failed to precipitated any $Ca(NO_3)_2$. Only when seeding the solution with fresh

$Ca(NO_3)_2$ crystals before cooling did precipitation take place. Analysis of the precipitates with SEM/EDS, Figure 8, revealed that the crystals were relatively pure and did not contain REEs that could be detected with EDS.

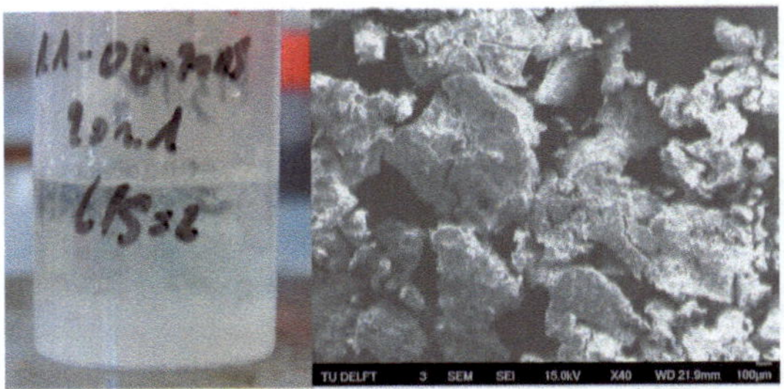

Figure 8: $Ca(NO_3)_2$ precipitation result and SEM

ICP-OES analysis of the remaining liquor shows that 65% of the Ca was removed through precipitation.

Proposed Recovery Process

Based on the results of this work a process for treating the apatite concentrate is proposed. Figure 9 shows the flowchart of the proposed process.

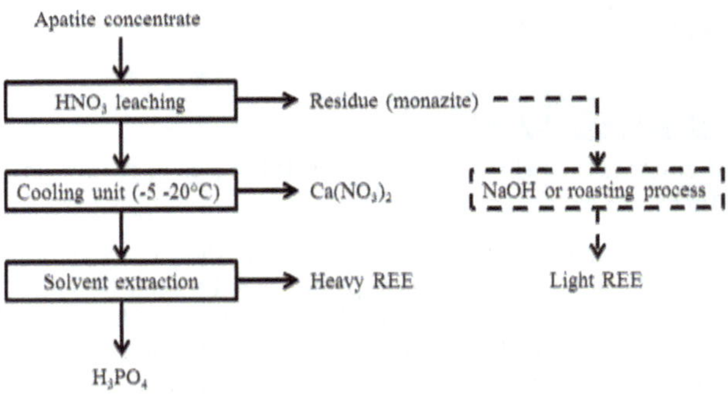

Figure 9: flowchart REE and P extraction process from apatite concentrate

The apatite concentrate is digested in concentrated HNO_3 to dissolve the apatite and heavy REEs. The liquor is then sent to a cooling unit, where the Ca is removed as $Ca(NO_3)_2$. The resulting solution is then processed using solvent extraction to separate the REEs and purify the H_3PO_4.

It is clear that acidic digestion under atmospheric condition fails to dissolve the monazite present in the apatite concentrate and that the monazite reports to the leach residue. As the value of the light REE is relatively low it is not economically interesting to process this residue further for its monazite. However should it become necessary to extract the light REEs as well as the heavy REEs, then the residue can be processed separately (as indicated by the dotted part of the flowchart in Figure 9). As the residue is essentially a monazite concentrate the technologies of the primary REE industry can be applied: NaOH digestion or roasting with $CaCl_2$ and $CaCO_3$ under a reducing and sulphidizing atmosphere [17].

Conclusions

Based on the leaching experiments performed in this work the following conclusions can be drawn:

(1) Full REE extraction proves impossible with single step HCl/HNO_3 leaching due to the refractory nature of the monazite. The monazite in the concentrate does not react and reports to the leaching residue.
(2) All apatite is digested and its phosphorous converted to H_3PO_4, which serves as the main by-product of the leaching process.
(3) The residue produced by the leaching reactions is minimal (1.7 wt.%) and contains the unreacted monazite, magnetite and refractories like magnesio-silicates. Other than the monazite this residue has no significant market value and can be safely disposed of.
(4) Yttrium and other heavier REEs, which are the more valuable REEs, show extraction ratios ranging from 75% to 100%, indicating that these REE are present in a phase other than monazite. These heavy REE are in all likelihood part of the apatite lattice as substitutional elements and can be efficiently recovered alongside the phosphorous of the apatite.
(5) Large excesses of acids are required to obtain REE extraction ratios above 20%. The excess acid maintains a sufficiently low pH to prevent any of dissolved REEs from re-precipitating.
(6) The REEs lost to the residue are mainly Ce and La, both REEs with currently no significant market value. The main value that is lost to the residue is 50% of the total Nd content. The residue can be post treated with NaOH or roasting to extract this lost Nd.
(7) Chilling the liquor to minus 20°C causes the precipitation of clean $Ca(NO_3)_2$ crystals. The precipitation requires seed crystals to initiate or the solution will stay super saturated.

Acknowledgements

The authors acknowledge the EU project FP7 REEcover (Project ID: 603564) for financial support. The authors would like to thank prof. Pålsson for insightful discussions and providing the material for this work, R. Hendrix and R. Huizenga for their aid with XRF and XRD analysis and H. de Waard for the ICP-MS analysis.

References

[1] K. Binnemans, P. T. Jones, B. Blanpain, T. Van Gerven, Y. Yang, A. Walton, and M. Buchert, "Recycling of rare earths: a critical review," *J. Clean. Prod.*, vol. 51, Jul. 2013, pp. 1–22

[2] L. Meyer and B. Bras, "Rare earth metal recycling," in *2011 IEEE International Symposium on Sustainable Systems and Technology (ISSST)*, 2011, pp. 1–6.

[3] "Critical Raw Materials - European Commission." [Online]. Available: http://ec.europa.eu/growth/sectors/raw-materials/specific-interest/critical/index_en.htm.

[4] G. B. Haxel, J. B. Hendrick, and G. J. Orris, "Rare Earth Elements—Critical Resources for High Technology," 2002.

[5] Y. Kanazawa and M. Kamitani, "Rare earth minerals and resources in the world," *J. Alloys Compd.*, vol. 408–412, 2006, pp. 1339–1343

[6] M. Humphries, "Rare Earth Elements: the global supply chain," Congressional Research Service, CRS Report for Congress, Dec. 2013.

[7] "REEcover." [Online]. Available: http://www.reecover.eu/.

[8] N. Krishnamurthy and C. K. Gupta, *Extractive metallurgy of rare earths*. CRC press, 2004.

[9] F. Habashi, "The recovery of the Lanthanides from phosphate rock," *J. Chem. Technol. Biotechnol.*, vol. 35A, 1985, pp. 5–14

[10] S. Peelman, Z. H. I. Sun, J. Sietsma, and Y. Yang, "Leaching of Rare Earth Elements: Past and Present," presented at the ERES2014: 1st European Rare Earth Resources Conference, Milos Greece, 2014, pp. 446–456.

[11] D. H. Krinsley, K. Pye, J. Sam Boggs, and N. K. Tovey, *Backscattered Scanning Electron Microscopy and Image Analysis of Sediments and Sedimentary Rocks*. Cambridge University Press, 1998

[12] B. Pålsson, O. Martinsson, C. Wanhainen, and A. Fredriksson, "Unlocking Rare Earth Elements from European apatite-iron ores," in *Proceedings of the 1st European Rare Earth Resources Conference (ERES2014), Milos (Greece), 4–7 September 2014*, 2014, pp. 211–220.

[13] B. S. Hopkins, *Chemistry of the rarer elements*. D.C. Heath and company, 1923.

[14] A. M. T. S. Bandara and G. Senanayake, "Leachability of rare-earth, calcium and minor metal ions from natural Fluorapatite in perchloric, hydrochloric, nitric and phosphoric acid solutions: Effect of proton activity and anion participation," *Hydrometallurgy*, vol. 153, Mar. 2015, pp. 179–189,

[15] D. Beltrami, G. J.-P. Deblonde, S. Bélair, and V. Weigel, "Recovery of yttrium and lanthanides from sulfate solutions with high concentration of iron and low rare earth content," *Hydrometallurgy (2015)*, http://dx.doi.org/10.1016/j.hydromet.2015.07.015

[16] F. T. Nielsson, *Manual of Fertilizer Processing*. New York: Marcel Dekker inc., 1987.

[17] R. R. Merritt, "High temperature methods for processing monazite: I. Reaction with calcium chloride and calcium carbonate," *J. Common Met.*, vol. 166, no. 2, Nov. 1990, pp. 197–210

Rare Metal Technology 2016
Edited by: Shafiq Alam, Hojong Kim, Neale R. Neelameggham, Takanari Ouchi, and Harald Oosterhof
TMS (The Minerals, Metals & Materials Society), 2016

FLUORINATION BEHAVIOR OF URANIUM AND ZIRCONIUM MIXTURE FOR FUEL DEBRIS TREATMENT

Nobuaki Sato[1], Akira Kirishima[1] and Tetsuo Fukasawa[2]

1: Institute of Multidisciplinary Research for Advanced Materials, Tohoku University, 2-1-1 Katahira, Aoba-ku, Sendai 980-8577, Japan
2: Hitachi-GE Nuclear Energy, 3-1-1, Saiwai-cho, Hitachi-shi, Ibaraki-ken, 317-0073, Japan

Keywords: Uranium, Zirconium, Fluorination, Fuel debris

ABSTRACT

To apply the fluoride volatility process for the treatment of fuel debris formed in the damaged nuclear reactors, study on the fluorination behavior of main constituents, such as fuel and cladding is essential. The mixture of uranium and zirconium oxides is formed under severe conditions such as high temperature and oxidizing atmospheres. In this paper, fluorination and volatilization behavior of uranium and zirconium oxides have been studied by TG-DTA method in the presence of fluorine. From the TG-DTA result of UO_2 in F_2 atmosphere, the weight gradually increased up to 7% at 350 °C followed by rapid weight decrease with exothermic peak showing the vaporization of uranium as UF_6. ZrO_2 showed similar fluorination behavior to that of UO_2. However, fluorination of UO_2-ZrO_2 seemed to be suppressed with increasing ZrO_2 amount. This may be caused by the formation of solid solution.

INTRODUCTION

The loss of coolant accident of Fukushima Daiichi NPS in 2011 caused the reactor core to melt resulting in the formation of fuel debris. The treatment of this fuel debris is scheduled to be within the next 20-25 years. By then methods for treating the fuel debris should be available. As one of such methods, application of the FLUOREX Process was considered. This process has been developed as pyro reprocessing process for spent nuclear fuel [1,2]. In this process, UO_2 fuel is fluorinated by fluorine forming UF_6. The UF_6 as well as PuF_6 vaporized and separated from non-volatile fission products (FPs). Furthermore, reaction of uranium oxides with fluorine and formation of uranium oxyfluoride were reported in our previous study [3]. The fuel debris consists of fuel materials and other components such as claddings, structural materials and concrete. In this debris, many phase consisting of oxides and metal seem to be co-existing. Therefore, information on the phase relation of fuel debris is needed first. And then, fluorination behavior of each component should be known. This project for the establishment of fuel debris fluorination treatment technology consists of four items, (1) fluorination of simulated fuel debris, (2) oxide conversion of residual fluorides, (3) thermodynamic evaluation, and (4) check and review by the external technical advisory committee. In the first item, fluorination for the pseudo binary and ternary compounds containing such as uranium, plutonium and zirconium, has been considered. Though no date was given for the fluorination of UO_2-PuO_2 system, the fluorination reaction could proceed by the experiments with actual spent fuel of high burn ups from the previous work. Other compounds like UO_2-Fe [4] and UO_2-ZrO_2 should be considered for their fluorination behavior in this project. In this paper, fluorination study is focused on the UO_2-ZrO_2 system. First, phase relation of the mixture of uranium and above zirconium oxides as a simulated fuel debris was analyzed by heat treatment under oxidative and reductive conditions. Then fluorination behavior the uranium and zirconium oxides was examined by TG-DTA method in the presence of fluorine.

EXPERIMENTAL

Materials used

The U_3O_8 was obtained by the oxidation of U metal turnings in air at 800 °C. Stoichiometric UO_2 was prepared by the H_2 reduction of U_3O_8 at 1000 °C. Zirconium metal (99.9%) and ZrO_2 (99.9%) were purchased from Wako Pure Chemicals Inc. and used without further purification. The mixed gas of He-5%F_2 was obtained from Central Glass Co. Ltd. The impurities are air < 0.02, CF_4 <0.002 and HF< 0.01 vol%. The 6N grade Ar gas was purchased from Nippon Sanso Co. Ltd. These gases were used without further purification.

Heat treatment

For the preparation of UO_2-ZrO_2 samples with UO_2:ZrO_2 ratios 9:1, 1:1 and 1:9, heat treatment of their mixture at 1200 °C was examined under reductive and oxidative conditions. This experiment was carried out by the horizontal type quartz reaction tube. The quartz boat containing the weighed amount of uranium and iron metal or oxide powders was set in the center part of the quartz reaction tube. After the evacuation and argon refill of the reaction system, the mixed gas of Ar+10%H_2 was introduced for reductive condition. In case of oxidative condition, Ar+2%O_2 gas was used. Then the electric furnace was heated up to the intended temperature at a heating rate of 10 °C/min. After the treatment, the phase relation of the product was analyzed by Rigaku Ultima-IV (40kV, 40mA) with Cu Kα radiation with a scan speed of 10 °/min.

Thermogravimetric measurement

Thermogravimetric measurements were conducted by using RIGAKU Thermoplus2 TG-DTA system. This system was reformed as an anti-corrosion type differential thermobalance. This thermobalance was installed in the high purity argon glove box for avoiding further corrosion by moisture. A sample of ca. 10 mg was weighed by electrical balance in a glove box and put in Ni sample pan followed by setting on the measuring bar of the thermobalance. After flashing Ar gas to the system, He-5%F_2 gas was introduced as well as Ar purge gas. The flow rates of He-5%F_2 and Ar gases were controlled by KOFLOC mass flow meter as 20 and 40 ml min^{-1}, respectively. Then the sample was heated to 500°C with a heating rate of 5°C min^{-1}. Since platinum components, such as sample pan and measuring bar are severely damaged over 500 °C in the presence of fluorine, all the experiments was conducted below this temperature. After the experiments, all the system was flushed with Ar to remove fluorine gas completely.

RESULTS AND DISCUSSION

Fluorination of Zr

In our previous study of fluorination of uranium oxides with fluorine [3], U_3O_8 reacts with F_2 forming volatile UF_6. This behavior is very similar to the case of UO_3. Here, TG-DTA measurement was conducted for zirconium metal and its oxides ZrO_2 for comparison first. Figure 1 shows TG-DTA curves for Zr metal with F_2. The TG curve gradually decreased with increasing temperature. The TG curve suddenly decreased with exothermic reaction around 400 °C showing the volatilization of some compounds. Finally, weight loss increased for 17 % at 500

°C. When Zr metal reacts with fluorine, ZrF_4 was formed with exothermic reaction. Since ZrF_4 is volatile at higher temperature, weight decrease appeared in this case.

$$Zr + 2F_2 = ZrF_4 \quad (1)$$

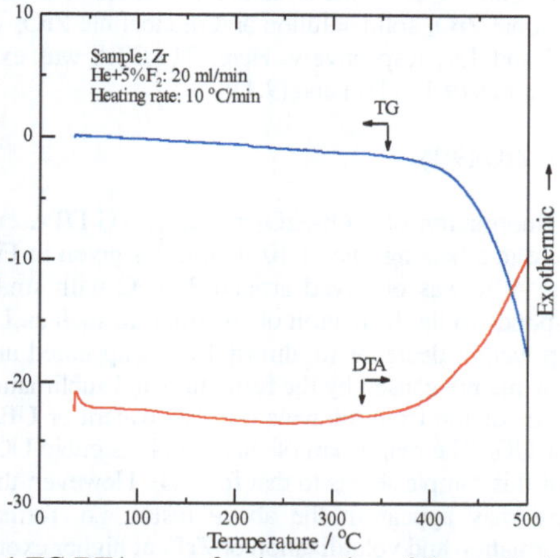

Figure 1 TG-DTA curve for the reaction of Zr metal with F_2.

Fluorination of ZrO_2

Figure 2 shows TG-DTA curve for ZrO_2 with F_2. The TG curve suddenly increased with exothermic reaction around 400 °C. Finally, weight increased to 17.5% at 500 °C. When ZrO_2 reacts with fluorine, oxyfluoride $ZrOF_2$ as given in eq. (2) seems to be formed.

$$2ZrO_2 + 2F_2 = 2ZrOF_2 + O_2 \quad (2)$$

The calculated weight increase of eq.(2) is 17.85% which is close to that obtained by TG. From the above results, ZrO_2 reacts with fluorine forming ZrF_4 via non-volatile $ZrOF_2$.

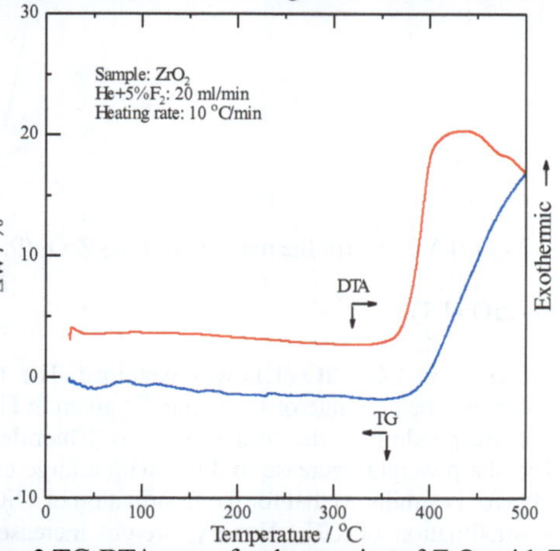

Figure 2 TG-DTA curve for the reaction of ZrO_2 with F_2.

Fluorination of UO_2+ZrO_2 mixture

According to the phase diagram for the UO_2-ZrO_2 system [5], a cubic UO_2 solid solution is formed in UO_2 rich region. On the other hand, ZrO_2 takes on the monoclinic structure at lower temperature and tetragonal one at higher temperature. From the XRD results the heat-treated UO_2-ZrO_2 samples, (1) cubic UO_2 solid solution, (2) cubic UO_2 and tetragonal ZrO_2 solid solutions, and (3) tetragonal ZrO_2 solid solution and monoclinic ZrO_2 were identified with UO_2-ZrO_2 ratios at 9:1, 5:5 and 1:9, respectively. Here, TG-DTA was examined for theUO_2-ZrO_2 mixture with UO_2:ZrO_2 ratios (9:1), (1:1) and (9:1).

a. Fluorination of UO_2-ZrO_2(9:1)

In the scheme of fluorination of UO_2-ZrO_2 mixture, TG-DTA curves for UO_2-ZrO_2 (9:1) sample in F_2 atmosphere at a heating rate of 10 °C min^{-1} is given in Figure 3. In this figure, the first weight increase to 4 % was observed around 350 °C with small exothermic peak. This weight increase corresponds to the formation of oxyfluoride such as UO_2F or $ZrOF_2$. After this phenomena, very steep weight decrease to almost 100 % appeared around 400°C with a large exothermic peak. This is mainly caused by the formation and sublimation of UF_6. In case of UO_2, weight increase by the oxidation from tetravalent to hexavalent or UF_4 formation do not appear before the formation of UF_6. The major part of this sample is cubic UO_2 solid solution. Thus the fluorination behavior of this sample seems to that for UO_2. However, the fluorination behavior of monoclinic ZrO_2 phase may appear in the above result; i.e. formation of $ZrOF_2$ at lower exothermic peak and formation and volatilization of ZrF_4 at higher exothermic one.

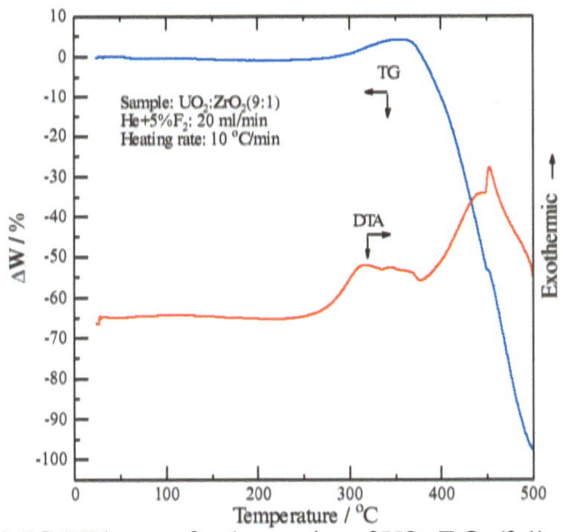

Figure 3 TG-DTA curve for the reaction of UO_2-ZrO_2 (9:1) with F_2.

b. Fluorination of UO_2-ZrO_2(1:1)

Next, fluorination behavior of UO_2-ZrO_2(1:1) was examined. The TG-DTA curves for UO_2-ZrO_2(1:1) in F_2 atmosphere at a heating rate of 10 °C $min^{-1}$is given in Figure 4. In this figure, the small weight increase corresponding to the formation of oxyfluoride such as UO_2F was not observed at 300 °C. The sharp weight decrease to 40 % with a large exothermic peak appeared around 400°C. This behavior is similar to that for the fluorination of U_3O_8 or UO_3. This is caused by the formation and sublimation of UF_6. Namely, weight increase by the oxidation from tetravalent to hexavalent or UF_4 formation do not appear before the formation of UF_6. However, half of the sample still remained without vaporization. Since this sample contains UO_2 solid solution and ZrO_2 solid solution with almost equal amount, only the former part was thought to be fluorinated resulting in the 50 % weight decrease. This result indicates that ZrO_2 solid solution part is very stable even in F_2 atmosphere.

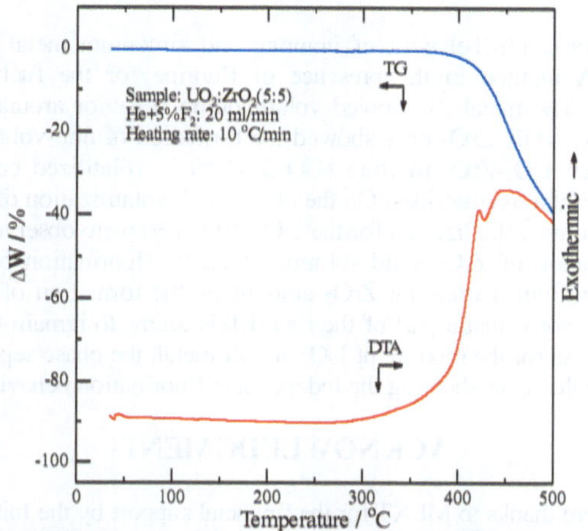

Figure 4 TG-DTA curve for the reaction of UO_2-ZrO_2 (5:5) with F_2.

c. Fluorination of UO_2-ZrO_2(1:9)

The TG-DTA curves for UO_2-ZrO_2(1:9) in F_2 atmosphere at a heating rate of 10 °C $min^{-1}$is given in Figure 5. There is no significant change in weight and heat. This means there is no reaction between UO_2-ZrO_2(1:9) sample and fluorine. This sample consists of monoclinic and tetragonal ZrO_2 solid solution phases from the result of XRD measurement. These results suggest that these ZrO_2 compounds are very stable even in F_2 atmosphere at high temperature.

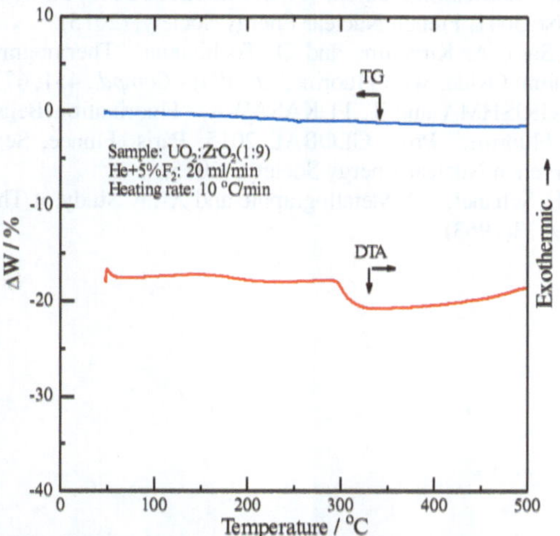

Figure 5 TG-DTA curve for the reaction of UO_2-ZrO_2 (9:1) with F_2.

CONCLUSION

In this study, fluorination behavior of uranium and zirconium metal and oxides has been studied by TG-DTA method in the presence of fluorine for the fuel debris treatment by FLUOREX process. The metal Zr showed volatilization behavior around 400 °C by forming volatile ZrF_4 directly, while ZrO_2 only showed the formation of non-volatile $ZrOF_2$ below 500 °C. The uranium rich UO_2-ZrO_2 mixture (UO_2:ZrO_2=9:1) volatilized completely by forming volatile UF_6 and ZrF_4 via oxyfluorides. On the other hand, volatilization of half part of the UO_2-ZrO_2(5:5) sample and no volatilization for the UO_2-ZrO_2(1:9) were observed. These results were caused by the formation of ZrO_2 solid solution. Namely, fluorination behavior of UO_2-ZrO_2 mixture is suppressed with increasing ZrO_2 amount by the formation of stable solid solution. This point should be noted that a part of the fuel debris seems to remain unreacted even by the fluorination method. As for the mixture of UO_2 and Zr metal, the phase separation between metal and oxide phases would occur showing the independent fluorination behavior.

ACKNOWLEDGMENT

The authors wish there thanks to MEXT for the financial support by the Initiatives for Atomic Energy Basic and Generic Strategic Research.

REFERENCES

1. Y. KANI, A. SASAHIRA, K. HOSHINO, and F. KAWAMURA, "New Reprocessing System for Spent Nuclear Reactor Fuel using Fluoride Volatility Method," *J. Fluor. Chem.*, 130, 74(2009)
2. T. FUKASAWA, A. SASAHIRA, K. HOSHINO, S. WATANABE, T. KIKUCHI, M. NAGATA, N. SATO and K. KIRISHIMA, "Development of fuel debris treatment technology by the fluorination method," Proc. GLOBAL 2015, Paris, France, September 21-24, 2015, Paper No.5041, French Nuclear Energy Society, (2015).
3. M. Komura, N. Sato, A. Kirishima and O. Tochiyama, "Thermogravimetric Study of the Reaction of Uranium Oxides with Fluorine," *J. Alloys Compd.*, 451, 673(2008).
4. N. SATO, A. KIRISHMA and T. FUKASAWA, "Fluorination Behavior of Uranium and Iron Oxides by Fluorine," Proc. GLOBAL 2015, Paris, France, September 21-24, 2015, Paper No.5370, French Nuclear Energy Society, (2015).
5. I. Cohen and B.E. Schaner, " A Metallographic and X-ray Study of The UO_2-ZrO_2 System," *J. Nucl. Mater.*, 9, 33(1963)

Rare Metal Technology 2016
Edited by: Shafiq Alam, Hojong Kim, Neale R. Neelameggham, Takanari Ouchi, and Harald Oosterhof
TMS (The Minerals, Metals & Materials Society), 2016

HYDROMETALLURGICAL RECOVERY OF RARE EARTH METALS FROM SPENT FCC CATALYSTS

M. Wenzel[1], K. Schnaars[1], N. Kelly[1], L. Götzke[1], S. Robles M.[2], K. Kretschmer[2], Phuc Nguyen Le[3], Dang Thanh Tung[3], Nguyen Huu Luong[3], Nguyen Anh Duc[3], Dang Van Sy[4], K. Gloe[1,*], and J. J. Weigand[1,*]

[1]TU Dresden, Department of Chemistry and Food Chemistry, 01062 Dresden, Germany
[2]Delta Engineering & Chemistry GmbH, 14476 Groß Glienicke, Germany
[3]Vietnam Petroleum Institute, VPI Power, Trung Kinh Street, Cau Giay District, Hanoi, Vietnam
[4]LILAMA EME, 41/168 Hao Nam Street, Dong Da District, Hanoi, Vietnam

Keywords: Rare earth, recovery, leaching, solvent extraction, FCC catalyst

Abstract

The recovery of rare earth metals from secondary sources has attracted much attention due to their ever expanding demand in the high-tech industry. The studies reported here focus on the hydrometallurgical recovery of lanthanum and cerium from spent fluid catalytic cracking (FCC) catalysts in a two-step process: leaching with nitric acid and solvent extraction by tri-*n*-butyl phosphate (TBP) and di(2-ethylhexyl)phosphoric acid (D2EHPA). The experiments show a high dissolution yield of about 93% lanthanum and 42% cerium in a single leaching step with 2 M (126 g/L) HNO_3 at 80 °C; only 11% aluminum has been dissolved simultaneously. In the subsequent solvent extraction step the best results for this leach liquor could be achieved using a 1:1 mixture of 25% (v/v) TBP (0.92 M) and 25% (v/v) D2EHPA (0.76 M) in n-decane without the need for any pH adjustment. In that case La(III) and Ce(III) can be extracted with 60% and 74% yield respectively in one stage from the majority of accompanying matrix elements. In particular no extraction of Al(III) could be observed under these conditions.

Introduction

In our high-tech society the comprehensive recovery and recycling of metals is a front-ranking requirement achieving both environmental and sustainable goals in metal winning [1-3]. That is of great importance especially also for the critical rare earth metals which are used for a variety of high-tech applications [4,5]. During the last years a lot of efforts have been taken to search for recovery and recycling possibilities of these critical metals from production scrap and residues, end-of-life products (phosphors, magnets, batteries, catalysts, polishing powders, glass) or landfilled residues [4,6-10] in order to guarantee the necessary supply chains now and in the future [11,12]. Despite the progress in the recycling methods for rare earth metals from different scrap and wastes only a minor amount is recycled today. The reasons for that range from economic over technological to societal factors [13]. The current intense research in this field will not only extend the recovery and recycling possibilities but also lead to interesting cost-efficient and sustainable new approaches [14-19]. In general, the mostly discussed concentration and purification methods for rare earth metals possess a hydrometallurgical basis using leaching, precipitation, solvent extraction and ion exchange process steps. These techniques are also well established in the mining industry of these metals [20,21]. A critical review on solvent extraction of rare earths was given in 2014 [22] discussing the influencing factors of different processes and the extractants used. Especially organophosphorus extractants or their synergistic mixtures play an important role and are widely used for this purpose [23,24]. Typical examples are tri-n-butylphosphate (TBP) as neutral and di(2-ethylhexyl)phosphoric acid (D2EHPA) as acidic complexing agents. Although several studies with different waste material are available in the literature it is necessary to prove the experimental conditions for every new application again. Differences or changes of the production conditions can influence the composition of the waste material and for this reason also subsequent recovery processes.

The aim of the present work consists in the development and testing of a recovery process of the rare earth components lanthanum and cerium from spent FCC catalyst material of a Vietnamese refinery of PetroVietnam. Based on the results for different industrial waste [4,7,9] in the literature a hydrometallurgical process including leaching, precipitation and solvent extraction steps was designed and tested in the lab. Here we will discuss the results of leaching studies of the spent FCC catalyst with HNO_3, HCl and H_2SO_4 as well as solvent extraction experiments using tri-n-butylphosphate, di(2-ethylhexyl)phosphoric acid and their mixtures. The formulas of both extractants are given in Figure 1 and Figure 2 depicts the spent catalyst.

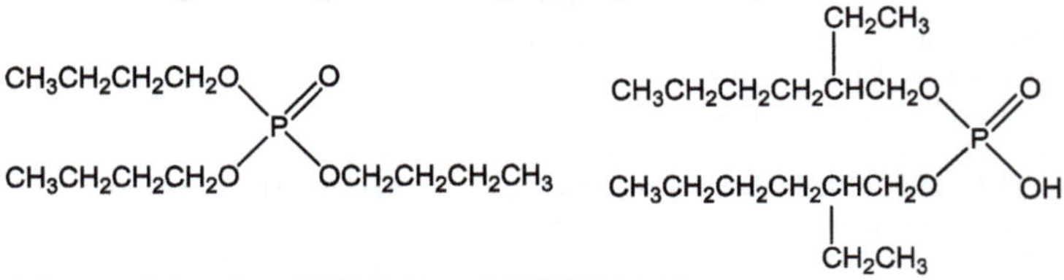

Figure 1: Structural formulas of TBP (left) and D2EHPA (right).

Figure 2: Spent FCC catalyst used in this study.

During our studies a similar approach for the recovery of La and Ce from spent FCC catalysts was published [25] which will be compared with the results of the present contribution.

Methodology

The powder X-ray diffraction was performed at 296(1) K on an X'Pert Pro MPD diffractometer (PANalytical) equipped with a curved Ge(111) monochromator by using Cu-$K_{\alpha1}$ radiation (λ=154.056 pm). The X-ray fluorescence (XRF) analysis of the spent catalyst was carried out by *CRB Analyse Service* GmbH Hardegsen using pressed tablets at an Axios WDXRF spectrometer (PANalytical); for the carbon analysis a vario Micro cube equipment (Elementar) was used. The particle size was determined on a Horiba LA950. All chemicals used were of analytical grade. The leaching experiments have been carried out in Erlenmeyer flasks (100 mL) using standard magnetic stirrer or mechanical overhead shaker (Heidolph). The solid residue was filtered through PTFE-filters with a pore size of 0.45 µm hydrophilized with methanol (Sartorius Stedim Biotech). A complete chemical digestion of the catalyst material was successful in a microwave MARS 5E (CEM) with a $HNO_3/HF/H_2O_2/H_3BO_3$ mixture. The resulting solution was analysed using an ICP-OES spectrometer OPTIMA 2000 DV (PerkinElmer). All extraction experiments were performed at room temperature in microcentrifuge tubes (2 mL) by mechanical shaking. The phase ratio $V_{(w)}:V_{(org)}$ was 1:1 (0.7 mL each); the shaking time was 60 min. Toluene or n-decane were used as organic diluent. All samples were centrifuged after extraction. The metal concentration in the aqueous phase was determined by ICP-OES analysis. The percentage extraction E in % was calculated from the experimentally determined distribution ratio D_M:

$$D_M = c_{M(org)}/c_{M(w)}$$
$$E[\%] = 100 \cdot D_M/(D_M+1).$$

Analysis of spent FCC catalyst

The spent FCC catalyst for this study was used without further treatment. The material consists of a modified zeolite Y [26] with a mean particle size of approximately 20 µm (Figure 3). The elemental analysis of the material shows a carbon content of 0.3 wt.% which points to a sufficient decoking of the FCC catalyst. The XRF analysis of the spent catalyst is summarised in Table 1. The highest concentration have been observed for aluminium (56.2 wt.% Al_2O_3; 29.74 wt.% Al) and silicon (39.8 wt.% SiO_2). The La_2O_3 content is about 1.5 wt.% (1.28 wt.% La) whereas Ce_2O_3 is present in 0.06 wt.% (0.05 wt.% Ce) of the total weight. The further accompanying metal oxides TiO_2, Fe_2O_3, NiO, V_2O_5, Co_2O_3, MgO, CaO, Na_2O and K_2O have concentrations lower than 1 wt.%. Furthermore the analytical data of a complete chemical digestion of the spent catalyst are given in Table 1. The comparison of these results shows that the two methods used lead to differing values especially for La_2O_3 (XRF 1.5 wt.%; ICP-OES 1.79 wt.%).

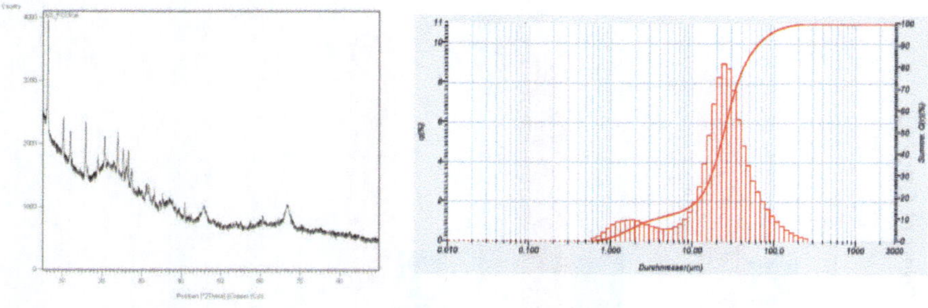

Figure 3: X-ray powder diffraction spectra (left) and particle size distribution (right) of the spent FCC catalyst used in this study.

Table 1: Comparison of the analyses of the spent FCC catalyst by XRF and of its solution after complete chemical digestion using ICP-OES.

Component	Content [wt.%]	
	XRF	Microwave / ICP-OES
Al_2O_3	56.2	56.13
La_2O_3	1.5	1.79
Ce_2O_3	0.059	0.09
Nd_2O_3	$<$0.025	0.02
SiO_2	39.8	nd
Fe_2O_3	0.78	0.98
TiO_2	0.96	0.43
V_2O_5	0.174	nd
CaO	0.21	0.38
MgO	0.17	0
NiO	0.449	0.2
Co_2O_3	0.042	0.01
Na_2O	0.306	0
K_2O	0.05	0

nd – not determined

Leaching Studies

A screening of the leaching power of three mineral acids, HCl, HNO_3 and H_2SO_4, were performed under comparable experimental conditions (T = 70 °C; t = 1 h) with 2 g of spent catalyst and 3 mL acid having a concentration of 140 g/L. The spent catalyst material was dried prior the leaching experiment for 48 h at 140 °C. The results are shown in Figure 4. In all three studied cases both the concentrations of rare earth ($c_{La(III)}$+$c_{Ce(III)}$ between 7.4 and 7.9 mg/mL) and matrix elements (between 19 and 22 mg/ml) in the leach liquor are very similar. This result

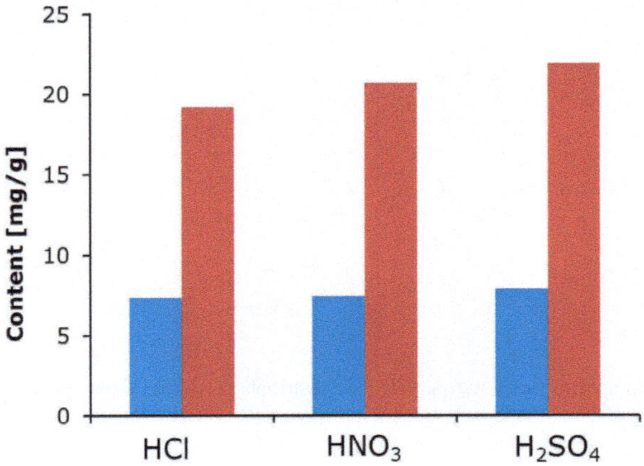

Figure 4: Results of leaching experiments (T = 70 °C; t = 1 h) for 2 g of the spent catalyst and 3 mL (140 g/L) of the acid demonstrated by the sum of rare earth metals (blue) and matrix elements (red) in the leach liquor.

points to the usability of all three acids for the leaching process which is in agreement with the literature [25,27-31]. Therefore the choice of the acid depends mainly from the requirements of the following concentration and separation steps of the recovery process. In the present work HNO_3 has been selected for leaching. In these studies the leach liquor results from the mixing of the filtrate and the washing solution filled up to 100 mL with water.

Detailed studies of the composition of the leach liquor were carried out after treatment of 10 g spent catalyst with 30 mL HNO_3 (126 g/L) at room temperature and a contact time of 1 hour. The results are shown for three independent experiments in Figure 5. The concentrations of La(III) and Al(III) are very similar with an average content of 96±4 mg/100 mL and 97±3 mg/100 mL, respectively. The content of Ce(III) was determined to be 2±0.1 mg/100 mL and for Nd(III) 0.04 ± 0.01 mg/100 mL, whereas the matrix elements with exception of Si(IV) (7±0.2 mg/100 mL) have values less than 4 mg/100 mL (Ca(II) 2±0.1 mg/100 mL; Co(II) < 1 mg/100mL; Fe(III) 2±0.1 mg/100 mL; K(I) < 1 mg/100 mL; Mg(II) 4±0.1 mg/100 mL; Na(I) 4±0.1 mg/100 mL; Ni(II) < 1 mg/100 mL, Ti(IV) < 1 mg/100 mL).

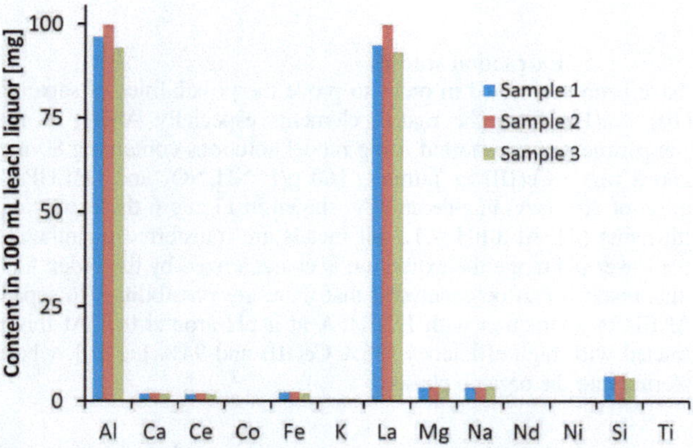

Figure 5: Metal content in 100 mL leach liquor after leaching of 10 g spent catalyst with 30 ml HNO_3 at 23 °C for 1 h.

In order to check the leaching yield in dependence on time, temperature, acid volume and concentration related experiments were done. The observed average concentrations of La(III), Al(III) and Ce(III) in the leach liquor are summarised for these influencing factors in Table 2.

It is clearly shown that there is only a negligible influence of the acid volume (column 1 and 3) and the acid concentration (column 1 and 4) on the leaching yield. However, a significant increase of the leached metal amount was observed both with increasing leaching time (column 1 and 2) and raising temperature (column 1 and 5). More than 90% of the La(III) amount present in the catalyst were dissolved in HNO_3 after 1 hour at 80 °C. In contrast to that only 11% of the matrix element Al(III) has been dissolved in the acid.

Table 2: Concentrations of Al(III), La(III) and Ce(III) in 100 mL leach liquor after leaching with HNO_3 in dependence on leaching time and temperature as well as acid volume and concentration.

Metal	Conditions and Content [mg/100 mL]				
	10 g catalyst 30 mL HNO_3 (126 g/L) 1 h 23 °C	10 g catalyst 30 mL HNO_3 (126 g/L) 6 h 23 °C	10 g catalyst 40 mL HNO_3 (126 g/L) 1 h 23 °C	10 g catalyst 30 mL HNO_3 (252 g/L) 1 h 23 °C	10 g catalyst 30 mL HNO_3 (126 g/L) 1 h 80 °C
Al(III)	97 ± 3	171 ± 4	98 ± 2	99 ± 8	336 ± 42
La(III)	96 ± 4	126 ± 2	96 ± 3	97 ± 7	142 ± 7
Ce(III)	2 ± 0.1	3 ± 0.1	2 ± 0.1	2 ± 0.2	3 ± 0.2

Therefore the concentration ratio between Al(III) and La(III) is reduced significantly from 20:1 in the starting material to approximately 2:1 in the leach liquor. This result considerably facilitates subsequent purification processes because the substantial proportion of Al(III) remains in the leaching residue.

Extraction studies

Solvent extraction studies have been performed in order to prove the possibilities of separation the rare earth metals La(III), Ce(III) from the matrix elements especially Al(III) as main component. The extraction experiments were started using model solutions containing 80 mg/L Al(III), 50 mg/L La(III) and 2 mg/L Ce(III) as nitrates, 160 g/L NH_4NO_3 and D2EHPA as extractant with a concentration of 50% (v/v) in n-decane. As shown in Figure 6 the extraction is increasing significantly with rising pH. At a pH = 1.7 all metals are transferred quantitatively into the organic phase. In the lower pH range the extraction is characterized by the order Al(III) < La(III) < Ce(III). From this result it can be concluded that there are possibilities to separate La(III) and Ce(III) from Al(III) by extraction with D2EHPA at a pH around 0.7. At this pH Ce(III) and La(III) are extracted with high efficiency [96% Ce(III) and 94% La(III)], whereas only 46% of Al(III) is transferred into the organic phase.

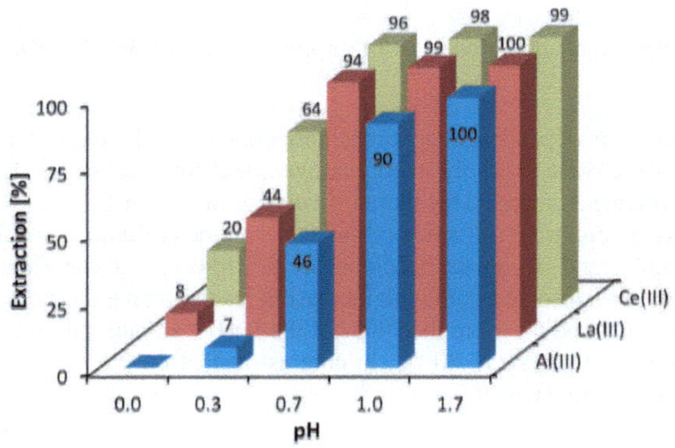

Figure 6: Extraction of Al(III), La(III), and Ce(III) by 50% (v/v) D2EHPA in n-decane in dependence on the pH of the aqueous phase: T = 23 °C, t = 1 h.

Further experiments using toluene as diluent or neutral phosphorus extractants, such as tri-*n*-butyl phosphate (TBP) or tri-n-octyl phosphine oxide (TOPO), do not show any improvement of the efficiency or selectivity of extraction for the interesting rare earth elements. However, applying a 1:1 mixture of TBP and D2EHPA as extractant the extraction of La(III) (87%) and Ce(III) (91%) remains in the high region, whereas the Al(III) extraction drops down from 46% to 8% (Figure 7). Thus, in case of the mixture the selectivity toward both rare earth metals is significantly increased and a separation of the two rare earth metals from Al(III) should be possible.

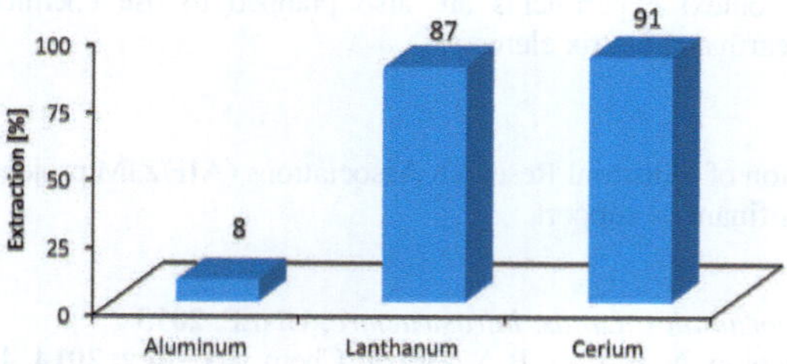

Figure 7: Extraction of Al(III), La(III) and Ce(III) by a mixture of 25% (v/v) TBP and 25% (v/v) D2EHPA in n-decane: pH = 1, NH_4NO_3 = 160 g/L, T = 23 °C, t = 1 h.

This conclusion has been proved by a solvent extraction experiment of a leach liquor obtained by acid leaching of 10 g spent catalyst with 30 mL HNO_3 (126 g/L) at 80 °C after 1 hour. The corresponding results for a 1:1 mixture of TBP/D2EHPA are given in Table 3. The results show a significant transfer of La(III) (66%, approximately 900 mg/L) and Ce(III) (73%, approximately 20 mg/L) into the organic phase, whereas Al(III) is not extracted at all. Also the matrix elements Si(IV), Mg(II) and Ca(II) are not extracted. Only Fe(III) is transferred into the organic phase quantitatively. There are several possibilities to remove the iron from the organic phase [32]. That will be investigated in near future in detail.

Table 3: Extraction of selected metals from a leach liquor by a mixture of 25% (v/v) TBP and 25% (v/v) D2EHPA in n-decane: T = 23 °C; t = 1 h; pH = 0.9.

Metal	Content [mg/L]		% extraction
	before extraction	after extraction	
Al	3976 ± 54	3994 ± 132	0
La	1482 ± 40	586 ± 23	60
Ce	27 ± 1	7 ± 1	73
Fe	86 ± 1	0.5 ± 2	99
Si	60 ± 1	61 ± 2	0
Ca	55 ± 1	57 ± 3	0
Mg	65 ± 2	66 ± 3	0

Conclusions

The study reported in this paper shall contribute to the development of a suitable hydrometallurgical process for the rare earth recovery from spent FCC catalysts. The results show that the combination of nitric acid leaching with subsequent solvent extraction using a TBP/D2EHPA mixture in n-decane as extractant allows a wide separation of the rare earth metals La(III) and Ce(III) from Al(III) and further accompanying metals. It is interesting to note that the leaching process can be operated in such a way that the necessary optimum pH for the solvent extraction step can be achieved without any modification of the feed solution. For the separation of the co-extracted Fe(III) from the organic phase in the extraction step several approaches are feasible which will be proved together with the detailed layout of the extraction/re-extraction step. In this context experiments are also planned to use chemical precipitation for the separation of rare earth and matrix elements.

Acknowledgement

The authors thank the German Federation of Industrial Research Associations (AIF/ZIM projects KF2807202RH3 and 3027402RH3) for financial support.

References

[1] M. Reuter, *Metal Recycling: Opportunities, Limits, Infrastructure*, UNEP, **2013**

[2] R. M. Izatt, S. R. Izatt, R. L. Bruening, N. E. Izatt, B.A. Moyer, Chem. Soc. Rev. **2014**, 43, 2451-2475

[3] T. Nakamura, K. Halada, Urban Mining Systems, Springer, Tokyo, **2014**

[4] D. Schüler, M. Buchert, R. Liu. S. Dittrich, C. Merz, *Study on Rare Earths and Their Recycling*, Öko-Institut, Darmstadt, **2011**

[5] T. Anand, B. Mishra, D. Apelian, B. Blanpain, JOM **2011**, 63, 8-9

[6] M. Tanaka, T. Oki, K. Koyama, H. Narita, T. Oishi, in Handbook on the Physics and Chemistry of Rare Earths, J.-G. Bünzli, V. Pecharsky, Eds., North-Holland, Amsterdam, **2013**, Vol. 43, Chapter 255, pp. 159-212

[7] K. Binnemans, P. T. Jones, B. Blanpain, T. V. Gerven, Y. Yang, A. Walton, M. Buchert, J. Clean. Prod. **2013**, 51, 1-22

[8] K. Binnemans, P. T. Jones, B. Blanpain, T. V. Gerven, Y. Pontikes, J. Clean. Prod. **2015**, 99, 17-38

[9] C. Tunsu, M. Petranikova, M. Gergoric, C. Ekberg, T. Retegan, Hydrometallurgy **2015**, 156, 239-258

[10] C. J. Ferron, P. Henry, in *Rare Earth Elements*, I. M. London, J. R. Goode, G. Moldoveanu, M. S. Rayat, Eds., Canad. Institute of Mining, Metallurgy and Petroleum, Westmont, **2013**, pp. 517-531

[11] A. Golev. M. Scott, P. D. Erskine, S. H. Ali, G.R. Ballantyne, Resour. Policy **2014**, 41, 52-59

[12] S. Massari, M. Ruberti, Resour. Policy **2013**, 38, 36-43

[13] T. E. Graedel, J:Allwood, J.-P. Birat, M. Buchert, C. Hagelüken, B. K. Reck, S. F. Sibley, G. Sonnemann, J. Ind. Ecol. **2011**, 15, 355-366

[14] T. V. Hoogerstraete, B. Blanpain, T. V. Gerven, K. Binnemans, RSC Adv. 2014, 64099-64111

[15] B. Van den Bogaert,D. Havaux, K. Binnemans, T. V. Gerven, Green Chem. **2015**, 17, 2180-2187

[16] S. Riano, K. Binnemans, Green Chem. **2015**, DOI: 10.10.39/c5gc00230c
[17] H. M. D. Bandara, K. D. Field, M. H. Emmert, Green Chem. **2015**, DOI: 10.10.39/c5gc01255d
[18] J. A. Bogart, C. A. Lippincott, P. J. Carroll, E. J. Schelter, Angew. Chem. Int. Ed. **2015**, 54, 8222-8225
[19] D. Kim, L. E. Powell, L. H. Delmau, E. S. Peterson, J. Herchenroeder, R. R. Bhave, Environ. Sci. Technol. **2015**, 49, 9452-9459
[20] C. G. Brown, L. G. Sherrington, J. Chem. Tech. Biotechnol. **1979**, 29, 193-209
[21] M. Cox, in *Solvent Extraction Principles and Practice*, J. Rydberg, M. Cox, C. Musikas, G. R. Choppin, Eds., Marcel Dekker, New York, **2004**, pp. 493-505
[22] F. Xie, T. A. Zhang, D. Dreisinger, F. Doyle. Miner. Eng. **2014**, 56, 10-28
[23] D. S. Flett, J. Organomet. Chem. **2005**, 690, 2426-2438
[24] D. Li, Y. Bai, X. Sun, S. Liu, in Rare Earth Elements, I. M. London, J. R. Goode, G. Moldoveanu, M. S. Rayat, Eds., Canad. Institute of Mining, Metallurgy and Petroleum, Westmont, **2013**, pp. 367-374
[25] V. Innocenzi, F. Ferella, I. De Michelis, F. Veglio, J. Ind. Eng. Chem. **2015**, 24, 92-97
[26] Zeolite database, International Zeolite Association, http://www.iza-structure.org/databases/
[27] B. Alexandre, N. Gerain, A. V. Lierde, in EMC'91: *Non-Ferrous Metallurgy-Present and Future*, J. Vereecken, Ed., Elsevier, London, **1991**, pp. 249-255
[28] Y. Jiang, A. Shibayama, K. Liu, T. Fujita, Hydrometallurgy **2005**, 76, 1-9
[29] L. V. Resende, C. A. Morais, Minerals Eng. 2010, 23, 277-280
[30] F. Yang, F. Kubota, Y. Baba, N. Kamiya, M. Goto, J. Hazard. Mater. **2013**, 254-255, 79-88
[31] X. Gao, W. T. Owens, US 0156116 A1 (**2012**)
[32] C. Xia, in *Rare Earth Elements*, I. M. London, J. R. Goode, G. Moldoveanu, M. S. Rayat, Eds., Canad. Institute of Mining, Metallurgy and Petroleum, Westmont, **2013**, pp. 255-275

Rare Metal Technology 2016
Edited by: Shafiq Alam, Hojong Kim, Neale R. Neelameggham, Takanari Ouchi, and Harald Oosterhof
TMS (The Minerals, Metals & Materials Society), 2016

DIRECT SOLVENT EXTRACTION OF NICKEL FROM SULFURIC ACID LEACH SOLUTIONS OF LOW GRADE AND COMPLICATED NICKEL RESOURCES USING A NOVEL EXTRACTANT OF HBL110

Li Zeng, Guiqing Zhang, Liansheng Xiao, Zuoying Cao

School of Metallurgy and Environment, Central South University; Changsha, 410083, China

Keywords: Nickel; Laterite; Electroplating sludge; Solvent extraction; Iron; Separation

Abstract

A new process of direct solvent extraction of nickel from sulfuric acid leach solutions of low grade and complicated nickel resources using a novel extractant HBL110 was proposed. The basic performance of new organic system was investigated and the continuous pilot plant and the industrial test were conducted. The results indicated that the extractant of HBL110 can be used to directly extract nickel selectively over iron, aluminum, manganese, magnesium, calcium and chromium from sulfuric acid leach solutions. The loaded organic solution can be easily stripped using dilute H_2SO_4 solutions. Both kinetics of extraction and stripping of nickel were fast. The continuous pilot plant test with the sulfuric acid leach solution of laterite containing 6.22g/L Ni^{2+}, 14.1g/L Fe^{3+}, 0.73g/L Fe^{2+}, 33.9g/L Mg^{2+}, 1.58g/L Al^{3+}, 0.495g/L Ca^{2+}, 0.97g/L Mn^{2+} and 0.5g/L Cr^{3+} as feed in mixer-settlers showed that the process had run stably with 98% recovery of nickel. The concentration of nickel in strip liquors reached as high as 50 g/L while the removals of Fe^{3+}, Fe^{2+}, Mg^{2+}, Al^{3+}, Ca^{2+}, Mn^{2+} and Cr^{3+} were 98.5%, 91.5%, 99.95%, 99.8%, 99.1%, 98.6% and 99.5%, respectively. The industrial test with the sulfuric acid leach solution of electroplating sludge containing 11.5g/L Ni^{2+}, 2.68g/L Fe^{3+}, 7.54g/L Cr^{3+}, 0.61g/L Mg^{2+}, 2.09g/L Al^{3+}, 0.68g/L Ca^{2+}, 0.022g/L Mn^{2+} and 0.154g/L SiO_2 as feed in mixer-settlers also showed that the process had operated successfully with 99.5% recovery of nickel. The concentration of nickel in strip liquors reached 42.7 g/L while the removals of Fe^{3+}, Cr^{3+}, Mg^{2+}, Al^{3+}, Ca^{2+}, Mn^{2+} and SiO_2 were 99.2%, 99.77%, 99.73%, 99.56%, 99.57%, 98.51% and 98.75%, respectively. The new process exhibited big advantages of short flow sheet, high recovery of nickel, high removal of impurities, low chemical reagent consumption and low cost. Application of this process in industry is expected to make important impact on the extraction of nickel from sulfuric acid leach solutions and to deliver good environmental and economical benefits.

Introduction

The current process on extraction of nickel from sulfuric acid leach solutions of laterite is "iron precipitation-nickel precipitation-acid dissolution-purification-electrowinning" [1]. The process has some drawbacks such as: a. long flowsheet; b. large loss of nickel, large energy consumption and large amount of residue during iron precipitation; c. low selectivity of nickel during nickel precipitation; d. Intermediate product obtained from nickel precipitation has to be dissolved by acid followed by further impurity removal using solvent extractant of P204 for purification, resulting in large cost.

Solvent extraction has been developed to explore the new process on selective separation of nickel directly over other metals, especially over iron[2~6]. Some of them have exhibited very good selectivity of nickel over other metals, but the difficult stripping of nickel and the fast degradation of extractant are the two main critical defects, which limited their successful application in industry.

In this study, a new and commercially available extractant of HBL110 was identified for the highly selective extraction of nickel over iron, aluminum, manganese, magnesium, calcium, zinc, chromium and silicon from sulfuric acid leach solutions, and a novel nickel extraction technology of direct solvent extraction (DSX) from sulfuric acid leach solutions of low grade and complicated nickel resources such as laterite and electroplating sludge was proposed.

Materials and method

Aqueous and organic solutions

The feed for metal extraction pH isotherm test was conducted using a synthetic solution to mimic the leach solution of laterite(Table 1). The feed for other tests were real leach solutions of laterite or electroplating sludge.

Table 1 The composition of synthetic leach solution

Elements	Ni	Co	Zn	Al	Fe	Mn	Mg	Ca	Cu	Cr
Concentration(g/L)	2.45	0.183	0.034	3.03	8.99	0.407	8.204	0.23	0.52	0.039

The reagent HBL110 was a new mixed extractant consisting of a pyridine carboxylate ester and a dinonylnaphthalenesulfonic acid, and provided by Hunan Hongbang New Materials Company, China and used as received without further purification.

Operation Method

The batch test of extraction and stripping was carried out by mixing equal volume (30 mL) each of aqueous solution and organic solution in 250 mL conical flask immersed in a water bath constant temperature oscillation for specified time. The shaking frequency was 180r/min and the temperature error was less than ± 0.1℃. After equilibrium, the phases were allowed to separate in separating funnels and their volume were measured. The initial pH and equilibrium pH of aqueous solutions were measured using a ROSS Sure Flow electrode (model 8172BN) coupled with a Hanna portable pH meter (model HI9025).

Chemical analysis

Inductively Coupled Plasma Atomic Emission Spectroscopy(ICP-AES) was used to detect the concentration of elements in aqueous solutions. The organic samples were stripped completely using 50 g/L H_2SO_4 solution and the concentration of elements in organic samples can be obtained by detecting it in stripping liquors.

Results and discussion

The metal extraction pH isotherm with HBL110 system

It can be seen from Fig. 1 that copper, nickel and cobalt could be highly selectively extracted over other metals such as iron, aluminum, manganese, magnesium, calcium and chromium in the tested pH range, realizing the direct separation of nickel over iron, aluminum, manganese, calcium and chromium. It was also observed that the extraction of nickel increased with the increase in equilibrium pH.

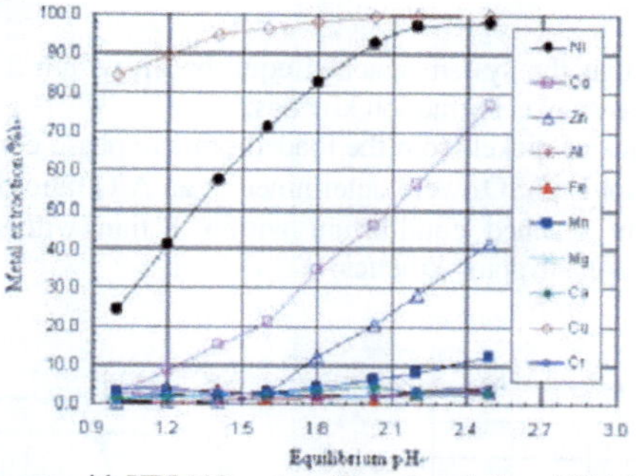

Fig.1 Metal extraction pH isotherms with HBL110 system (Organic solution: 45%(v/v) HBL110+ 55%(v/v) sulfonated kerosene)

<u>Characterization of HBL110 system</u>

Nickel extraction and stripping distribution isotherm

The nickel extraction distribution isotherm was determined using 45%(v/v) HBL110 in sulfonated kerosene and a real acid leach solution of laterite containing 3.7g/L Ni at pH2.0 and 25℃(Fig. 2). As seen, the saturated extraction capacity of nickel with the HBL110 system was 5.0 g/L. The nickel stripping distribution isotherm was determined with the loaded organic phase containing 4.88 g/L Ni and a strip solution of 0.5 mol/L H_2SO_4 at 30℃(Fig. 3). It can be estimated from Fig. 2 and Fig. 3 that the HBL110 system can be used to extract nickel from the acid leach solution completely in a few stages of counter-current extraction and the nickel in the loaded organic phase can be easily stripped by dilute H_2SO_4 solution.

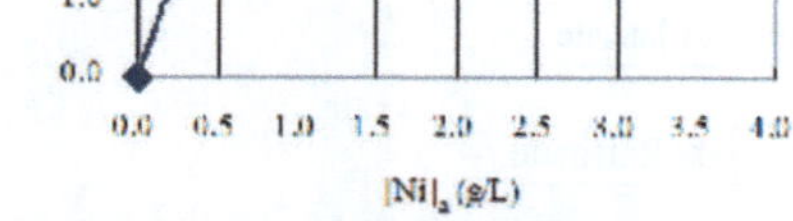

Fig. 2 Nickel extraction distribution isotherm

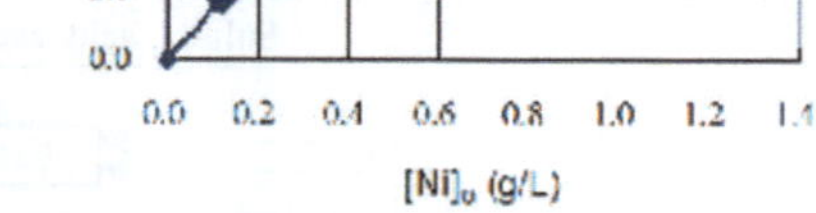

Fig. 3 Nickel extraction distribution isotherm

Nickel extraction and stripping kinetics

The extraction kinetics of nickel with the 45%(v/v) HBL110 system in sulfonated kerosene and the real acid leach solution containing 3.7g/L Ni at an A/O ratio of 1 and 25℃ is shown in

Fig. 4. It was found that the system reached equilibrium within 3 min with over 80% nickel extracted, indicating fast nickel extraction kinetics.

The stripping kinetics of nickel from the loaded organic phase containing 4.66 g/L Ni using a strip solution of 0.5 mol/L H_2SO_4 were determined at an A/O ratio of 1 and 30℃(Fig. 5). It was found that the system reached equilibrium within 1 min with over 90% nickel stripped, indicating very fast nickel stripping kinetics.

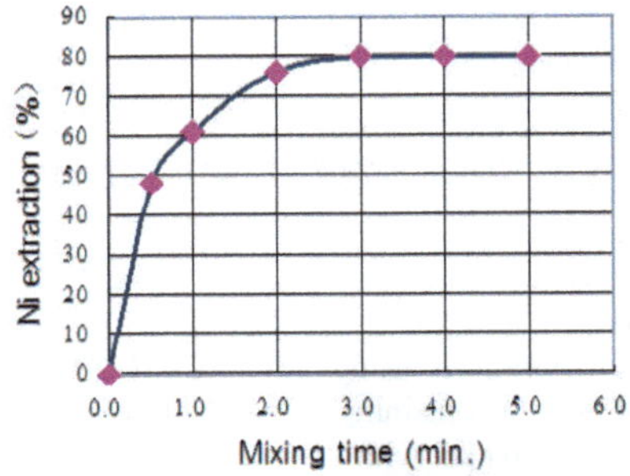

Fig. 4 Nickel extraction kinetics

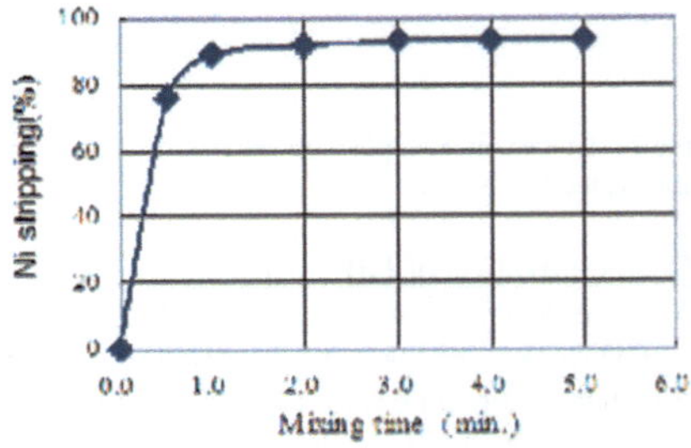

Fig. 5 Nickel stripping kinetics

<u>Fully-continuous pilot plant campaigns</u>

The principle flowsheet of nickel extraction from acid leach solutions of laterite for the pilot plant test is shown in Fig. 6. The system of multi-stage mixer settlers (with individual effective volumes of 10L for mixer) was used in pilot plant campaigns. Two types of acid leach solutions including heap leach solutions of magnesian laterite and agitation leach solutions of ferruginous laterite were treated in pilot plant campaigns. The system included 5 stages of extraction, 1 stage of scrubbing, 1 stage of settling of organic phase, 4 stages of stripping, 3 stages of washing after stripping and 1 stage saponification for heap leach solutions of magnesian laterite, while the stage of scrubbing was cancelled for agitation leach solutions of ferruginous laterite. The pilot plant test was operated for over 360 h with heap leach solutions of magnesian laterite and for over 288 h with agitation leach solutions of ferruginous laterite.

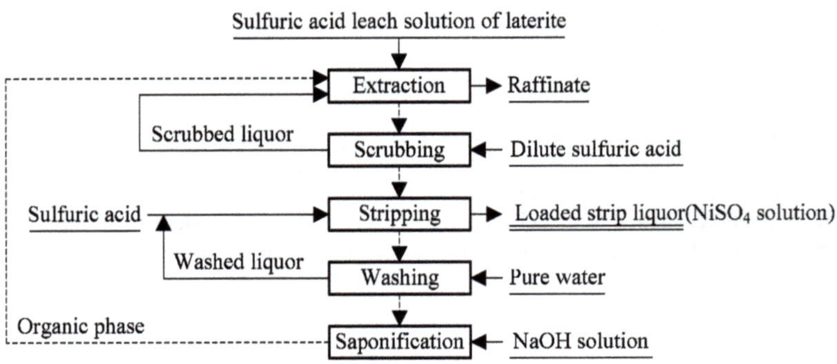

Fig. 6 Novel process on the extraction of Ni from sulphuric acid leach solution of laterite

The fully-continuous pilot plant campaigns showed that the process was very stable and easy to control. The typical results in steady state are shown in Tables 2 and 3.

Table 2 Fully-continuous pilot plant test results with heap leach solution

Element(g/L)	Ni	Co	Cu	Fe^{3+}	Fe^{2+}	Mg	Mn	Ca	Zn	Al	Cr	SiO_2
Feed	3.75	0.057	0.005	2.92	1.35	42.3	0.31	0.47	0.016	0.93	0.096	0.28
Strip liquor	39.14	0.184	0.018	0.30	0.65	0.28	0.015	0.0058	0.0059	0.013	0.003	0.010
Raffinate	0.06	0.03	0.0005	2.21	1.24	37.9	0.32	0.50	0.013	0.82	0.13	0.18
Scrub liquor	1.13	0.043	---	0.21	0.62	1.88	---	0.022	---	---	---	---
Extraction(%)	**98.2**	**42.1**	**89.0**	---	--	--	--	--	--	--	--	
Impurity removal(%)	---	---	---	**99.0**	**95.4**	**99.9**	**99.5**	**99.9**	**96.5**	**99.9**	**99.7**	**99.7**

It can be seen from Table 2 that for the heap leach solution of magnesian laterite, the average extraction of nickel reached 98%, leaving 0.06 g/L Ni in the raffinate. Around 40% of cobalt and 90% of copper were co-extracted which can be separated from nickel in the loaded strip liquor. The removals of impurity such as iron(III), aluminum, magnesium, manganese and calcium were up to 99% except 95% for iron(II) and 96.5% for zinc, indicating most of impurities can be almost completely removed. The concentration of nickel in strip liquors nearly reached 40 g/L, which was 10 times of that in feed.

With the agitation acid leach solutions of ferruginous laterite, it can be seen from Table 3 that the results were quite similar with those for heap leach solutions of magnesian laterite. The average extraction of nickel reached 99.5% and the concentration of nickel in strip liquors nearly reached 50g/L. More than 60% of cobalt and 96% of copper were co-extracted. The removals of impurity were still as high as 99%. It is noted that the extractions of nickel, cobalt and copper were a little higher than those with heap leach solutions, and the removals of impurity were a bit lower than those with heap leach solutions. This is because that the step of scrubbing was cancelled.

The most impurity contents in strip liquors with two acid leach solutions were much less compared with those in intermediate product of crude $Ni(OH)_2$ produced with traditional method(Table 4).

Table 3 Fully-continuous pilot plant test results with pressure acid leach solution

Element(g/L)	Ni	Co	Cu	Fe^{3+}	Fe^{2+}	Mg	Mn	Ca	Zn	Al	Cr	SiO_2
Feed	6.22	0.173	0.012	14.13	0.73	33.94	0.97	0.495	0.035	1.58	0.50	0.133
Strip liquor	49.58	0.672	0.070	1.690	0.495	0.90	0.11	0.035	0.018	0.032	0.021	0.021
Raffinate	0.07	0.06	0.0004	---	---	29.40	0.77	0.38	0.030	1.34	0.34	0.080
Extraction(%)	**99.50**	**61.85**	**96.33**	---	--	--	--	--	--	--	--	

Impurity removal(%)	---	---	---	**98.50**	**91.49**	**99.95**	**98.6**	**99.11**	**93.62**	**99.75**	**99.47**	**98.07**

Table 4 Typical components of intermediate product $Ni(OH)_2$ with traditional method

Element	Ni	Mg	Al	Si	Ca	Cr	Mn	Fe	Co	Cu	Zn
Crude $Ni(OH)_2$	48.0	3.07	0.08	0.44	0.05	0.013	2.46	0.26	2.15	0.05	0.32

Industrial test with sulfuric acid leach solutions of electroplating sludge

The new process of direct solvent extraction of nickel from acid leach solutions of electroplating sludge has been successfully applied in a domestic plant for 800 tons $NiSO_4$/a production since from August, 2014. The system of multi-stage mixer settlers (with individual effective volumes of 500L for mixer) was used in industrial production. The system included 5 stages of extraction, 1 stage of settling of organic phase, 4 stages of stripping, 3 stages of washing after stripping, 1 stage of settling after washing and 1 stage of saponification. The system has been operated for over 12 months and the results showed that the process was very stable and easy to control. The typical results are shown in Table 5.

Table 5 Typical results of industrial application of DSX on extraction of nickel from acid leach solutions of electroplating sludge

Element(g/L)	Ni	Fe	Mg	Mn	Ca	Zn	Al	Cr	SiO_2
Feed	11.5	2.68	0.61	0.022	0.68	2.30	2.09	7.54	0.154
Strip liquor	42.73	0.079	0.0061	0.0012	0.011	0.155	0.034	0.064	0.0071
Raffinate	0.016	2.25	0.58	0.021	0.63	2.28	2.01	6.94	0.145
Extraction(%)	**99.5**	---	--	--	--	--	--	--	
Impurity removal(%)	---	**99.2**	**99.73**	**98.51**	**99.57**	**98.18**	**99.56**	**99.77**	**98.75**

It can be seen from Table 5 that the average extraction of nickel reached 99.5%, leaving 0.016 g/L Ni in the raffinate. The removals of impurity such as iron, magnesium, manganese, calcium, zinc, aluminum, chromium and silicon were almost around 99%, indicating very good selectivity of nickel over these impurities. The concentration of nickel in strip liquors reached 40 g/L. The successful industrial application of the new DSX process in the extraction of nickel from acid leach solutions of electroplating sludge demonstrated its reliability and technical advancement.

Based on the investigation and discussion above, it can be concluded that nickel can be directly extracted and completely separated from iron, aluminum, manganese, magnesium, calcium, chromium and silicon in the new DSX process with novel extractant HBL110. The loaded strip liquor with high concentration of nickel and a small amount of impurities can be used to produce the final product of nickel sulfate or nickel metal after further fine purification.

Conclusion

The new and commercially available extractant of HBL110 can be used to directly extract nickel over iron, aluminum, calcium, magnesium, manganese, chromium and silicon from acid leach solutions of low grade and complicated nickel resources such as laterite and electroplating sludge. The pilot plant and the industrial test results demonstrated that the novel process of direct solvent extraction (DSX) of nickel exhibited big advantages of short flow sheet, high recovery of nickel and high removal of impurities, low energy and chemical reagent consumption, resulting in a large decrease of cost compared to that with traditional process. Application of this DSX process in industry is expected to make very important impact on the extraction of nickel from sulfuric acid leach solutions of low grade and complicated nickel resources and to deliver good environmental and economical benefits.

References

[1] Liu, D.X. Recent development in technology on recovery of nickel and cobalt from laterite. Nonferrous Metals, 2002,3: 6-12.

[2] Flett, D.S. Cobalt-nickel Separation in Hydrometallurgy: A Review. Chemistry for Sustainable Development 2004, 12, 81-91.

[3] Ritcey, G.M. Processes. In Solvent Extraction: Principles and Application to Process Metallurgy; Revised 2nd edition, 2006. Published by G.M. Ritcey and Associates Incorporated, Ottawa; Vol. 2, 356-359.

[4] Ritcey, G.M. Theory and Practice. In Solvent Extraction: Principles and Application to Process Metallurgy. Revised 2nd edition, 2006. Published by G.M. Ritcey and Associates Incorporated, Ottawa; Vol. 1, 56-57.

[5] Preston, J.S. Recent Development in the Separation of Nickel and Cobalt from Sulphate Solutions by Solvent Extraction. J. of South African IMM, June, 1983, 126-132.

[6] Cheng, C.Y., Barnard, K.R., Zhang, W.S. and Robinson, D.J. Synergistic solvent extraction of nickel and cobalt: A review of recent developments. Solvent Extraction and Ion exchange, 2011, 29:719-754.

Rare Metal Technology 2016
Edited by: Shafiq Alam, Hojong Kim, Neale R. Neelameggham, Takanari Ouchi, and Harald Oosterhof
TMS (The Minerals, Metals & Materials Society), 2016

ACID LEACHING ZINC AND INDIUM WITH REDUCTION FERRIC SIMULTANEOUSLY FROM MARMATITE AND HIGH-IRON NEUTRAL LEACHING RESIDUE

Zhigan Deng[1], Fan Zhang[1], Chang Wei[1], Cunxiong Li[1], Xingbin Li[1], Gang Fan[1], Minting Li[1]

[1]Faculty of Metallurgical and Energy Engineering, Kunming University of Science and Technology; No.253 Xuefu Road; Kunming, Yunnan, 650093, China

Keywords: Zinc Concentrate of High-iron; Neutral Leaching Residue; Simultaneous Leaching; indium; Reduction of Ferric

Abstract

Marmatite is an important resource of zinc, which exist in forms of high iron and high indium. In order to recycle the valuable metal from these ores, the testwork of simultaneous leaching on neutral residue mixed with zinc concentrate of high iron and high indium has been conducted. The primary objectives are to leach the zinc and indium remaining in the neutral leach residue as ferrite using acid, and to reduce the ferric iron in solution to the ferrous by zinc concentrate, thereby extracting the zinc and indium in the zinc concentrate. The results indicated that in the following conditions, mass ratio of neutral residue to zinc concentrate of 1:0.25, particle size of 74-58 μm, initial sulfuric acid concentration of 150 g/L, liquid to solid ratio of 8 L/kg, temperature of 90°C, reaction time of 4h, the leaching efficiency of zinc and indium were over 95%, nearly 90% Fe was Ferrous in the leaching solution. This process effectively extracted zinc and indium and converted Fe^{3+} into Fe^{2+} in solution, and simplified the traditional process of reduction after leaching.

1. Introduction

Indium is used increasingly in the electronic industry as a doping agent. Minerals, rich in indium, are few and very rare [1, 2]. The indium is mostly associated in trace amounts in zinc ores such as sphalerite, and it has therefore been recovered as a by-product of zinc [3, 4].
Marmatite is an important resource of zinc, which exist in forms of high iron and high indium. The conventional zinc recovery process is the roast-leaching-electrowinning process [5]. In this process, a significant part of zinc ferrite is obtain when zinc concentrate was roasted. The indium what is associated in zinc concentrate enter into the crystal lattice of zinc ferrite through substituted iron to form indium-bearing zinc ferrite [6]. In the hydrometallurgical zinc process to treat zinc concentrate, the majority of indium comes into the leach residue which has a significant part of indium-bearing zinc ferrite. Unfortunately, the indium-bearing zinc ferrite is stable and difficult to dissolve in low acid. Hot acid leaching is the main way to dissolute zinc ferrite [7-9]. There is much ferric ion in lixivium of the hot acid leaching. Sphalerite concentrate can successfully be used just as a reductant in acidic solution [10-14]. The reduction of ferric to ferrous, and then oxidation and precipitation of ferrous ion as hematite would solve the problem with losses of zinc in jarosite process; and also overcome the problems in the conversion process,

where several valuable metals (Ag, Au, Pb) are lost in the peripitation. What is more, it is good for separations of indium and zinc when the ferric ion is converted to the ferrous ion [15, 16]. Using sphalerite concentrate as reductant in simultaneous leaching, the majority of the ferric ion present in the leach process is converted to the ferrous ion. The main reactions are as follows:

$$ZnO\cdot Fe_2O_3+4H_2SO_4=ZnSO_4+Fe_2(SO_4)_3+4H_2O \quad (1)$$

$$ZnO\cdot Fe_2O_3+ZnS+4H_2SO_4=2\ FeSO_4+2ZnSO_4+S^0+4H_2O \quad (2)$$

In order to recycle the valuable metal from these ores, the testwork of simultaneous leaching on neutral residue mixed with zinc concentrate of high iron and high indium has been conducted. The aim of this research was to state the simultaneous leaching process of zinc residue what had a significant part of zinc ferrite by sulfuric acid in the presence of high-iron sphalerit as reductant. The extraction of indium and zinc from residue was demonstrated. The effects of high-iron sphalerite, sulfuric acid concentration, temperature, particle size and time were investigated.

2. Experimental

2.1. Materials

The zinc residue and the high-iron sphalerite used in the present were from YunNan Province, China. Analytical grade reagents, including sulfuric acid, were purchased from AR, Chengdu Union Institute Chemical & Reagent.

2.2. Experimental procedure

The samples were crushed and ball milled to 85% < 0.074 mm and dried at 65°C in an oven for 24 h. Leaching experiments were conducted at atmospheric pressure in an automation cylinder reactor with 10 L.

At the end of experiment, the ore pulp was withdrawn and filtered. The solid residue and solution were chemically analyzed. The extractions of Zn, In and Fe were calculated.

2.3. Analytical methods

Zinc was analyzed by complex titration with EDTA. Indium was analyzed by atomic absorption spectrometry (Shanghai Morning Field Equipment Co., Ltd). Fe was analyzed by complex titration with potassium bichromate. X-ray powder diffraction (XRD) was carried out using Rigaku D/MAX 2500v diffractometer (Japan). The potential of solution was measured using a PHSJ-4F digital pH meter by a platinum electrode with a Ag/AgCl electrode used as the reference electrode (Shanghai Leici Instrument Co., Ltd).

3. Results and Discussion

3.1. Characterization of the materials

The zinc residue and the high-iron sphalerite used in the present were from YunNan Province, China. The compositions of the zinc residue and the high-iron sphalerite were listed in Table I. X-ray diffraction of zinc residue indentified zinc ferrite ($ZnFe_2O_4$) as the main mineral components in Figure 1. No special indium bearing minerals were identified. It can be seen from

Figure 2 that the main compositions of the high-iron sphalerite were sphalerite, chalcopyrite and herzenbergite.

Table I. The main chemical component of materials (wt.%)

	Zn	Fe	In	S	SiO_2	Ag	Cu	Pb	Sn	CaO	MgO
Residue	24.75	25.68	566g/t	3.75	8.68	97.2g/t	1.12	0.099	0.21	2.68	1.94
Concentrate	44.38	15.89	358g/t	31.89	4.45	34.5	0.58	0.061	0.18	0.58	2.95

	Co	Ni	Cd	Al_2O_3	Mn	As	Sb	F	Cl	Ge	Ga
Residue	0.0048	0.0065	0.087	1.07	0.13	0.48	0.004	0.036	0.01	trace	trace
Concentrate	0.0061	0.0031	0.16	0.63	0.06	0.33	0.0004	0.2	0.01	trace	trace

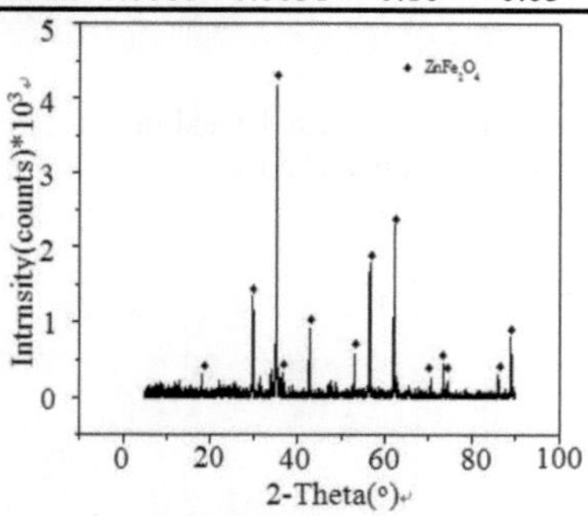

Figure 1. XRD of the zinc residue.

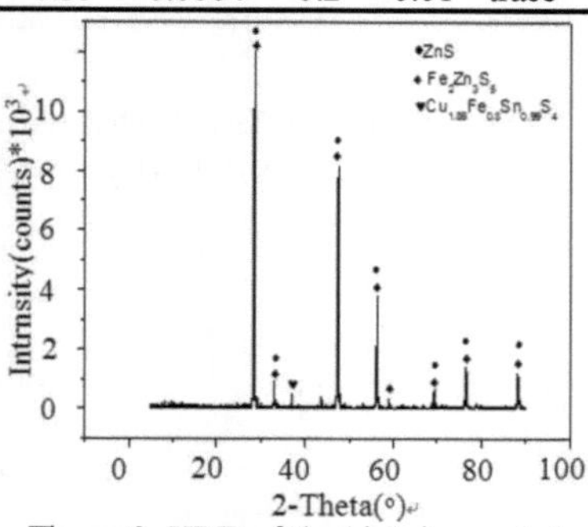

Figure 2. XRD of the high-iron sphalerite.

3.2 Effort of high-iron sphalerite addition

Addition of high-iron sphalerite into the reactor provided a reductant of converting Fe^{3+} into Fe^{2+}, thereby dissolving iron into the solution. Figure 3 presented the effect of high-iron sphalerite on extractions of Zn, Fe and In. It was worth noting in Figure 3 that the extractions of metal were not belonged to only the zinc residue, but also the high-iron sphalerite when the high-iron sphalerite was added. It was showed in Figure 3 that the extractions were ~88% zinc, ~78% iron, ~81% indium after 5 h without high-iron sphalerite. With the addition amount of high-iron sphalerite was 0.9 times of theoretical value, the extractions of Zn, Fe and In were increased significantly giving ~96%, ~93% and ~95%, respectively. Unfortunately, the extractions of Zn, Fe and In had stopped the ongoing growth when the addition amount of high-iron sphalerite was over 0.9.

Figure 4 presented the effect of high-iron sphalerite on concentrations of Fe^{3+} and Fe^{2+} in lixivium of the simultaneous leaching. It can be seen from Figure 4 that the Fe^{3+} was decreased and the Fe^{2+} was increased with the addition amount of high-iron sphalerite increased. If Fe^{2+} is present in the lixivium, in the absence of other reductants it becomes the ion determining the redox potential. With the Fe^{2+} increased, the redox potential decreased.

Figure 5 presented the effect of high-iron sphalerite on reacted acid and redox potential in the lixivium. It can be seen from Figure 5 that the redox potential was decreased with the addition amount of high-iron sphalerite increased.

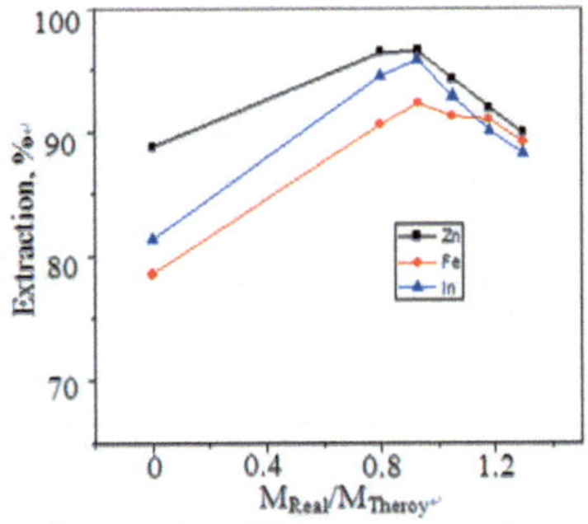

Figure 3. Effect of sphalerite addition on extractions of Zn, Fe and In. (90°C,150 g/L H_2SO_4, particle size: 74-58 μm and 5 h).

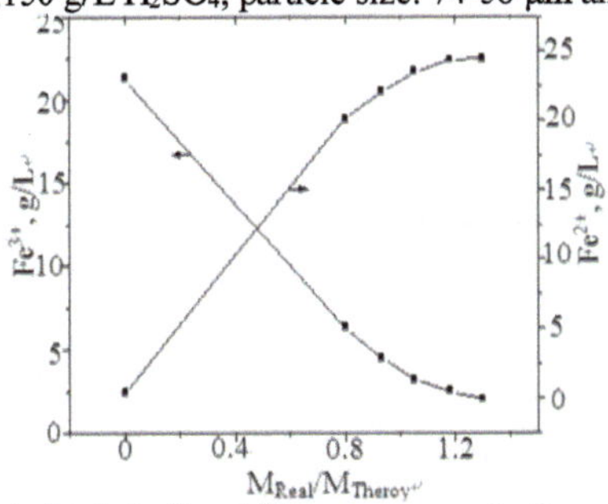

Figure 4. Effect of sphalerite addition on concentration of $Fe^{3+}$and Fe^{2+}. (90°C,150 g/L H_2SO_4, particle size: 74-58 μm and 5 h).

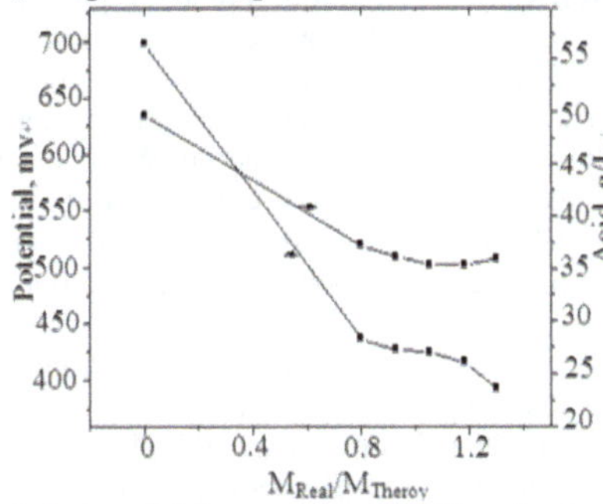

Figure 5. Effect of sphalerite addition on reacted acid and redox potential of lixivium. (90°C,150 g/L H_2SO_4, particle size: 74-58 μm and 5 h).

3.3 Effect of sulfuric acid concentration

In the simultaneous leaching process, the reacted acid of the lixivium improved with increasing in sulfuric acid concentration, and selecting the appropriate concentration of sulfuric acid was necessary. The effect of sulfuric acid concentration on the leaching of metals was studied in the range 90 g/L to 180 g/L. Figure 6 showed the extractions of Zn, Fe and In as a function of sulfuric acid concentration in leaching solution. It can be seen from Figure 6 that the curves of leaching efficiencies of Zn, Fe and increased faster with the increased in the sulfuric acid concentration from 90 to 150 g/L, but leaching efficiencies of Zn, Fe and In were slightly improved when the concentration was over 150 g/L. The leaching efficiencies of Zn, Fe and In reached ~96%, ~93% and ~95% when sulfuric acid concentration was 150 g/L.

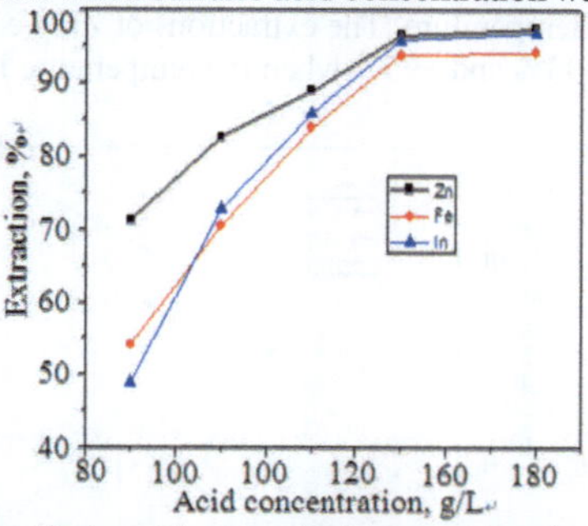

Figure 6. Effect of sulfuric acid concentration on extractions of Zn, Fe and In.
(The addition of sphalerite 0.9, 90°C, particle size: 74-58 μm and 5 h).

3.4 Effect of time

The effect of reactive time on the extractions of metal was studied in the range of 60 min to 300 min. It was showed in Figure 7 that the extractions of Zn, Fe and In increased from ~81%, ~65% and 59% to ~96%, ~92% and ~94% when the time increased from 60 min to 240 min, respectively. But a further increase in time to 300 min had no significant affect.

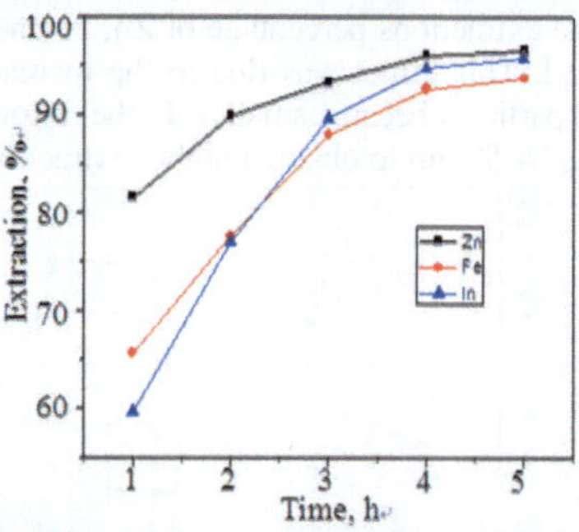

Figure 7. Effect of time on extractions of Zn, Fe and In.
(The addition of sphalerite 0.9, 90°C, 150 g/L H_2SO_4 and particle size: 74-58 μm)

3.5 Effect of temperature

The extractions rate of metal were also affected by leaching temperature. High temperature was good to the dissolutions rate of zinc ferrite and sphalerite. However, high temperatures had disadvantages such as higher energy consumption and higher evaporation. A medium temperature range from 60 to 90°C was selected and its affect on metal extractions was studied. Figure 8 showed the effect of temperature on the extractions of metal. The extractions of metal increased with increasing of temperature. The extractions of Zn, Fe and In increased from ~69%, ~65% and ~63% to ~96%, ~93% and ~95% when the temperature increased from 60°C to 90°C, respectively.

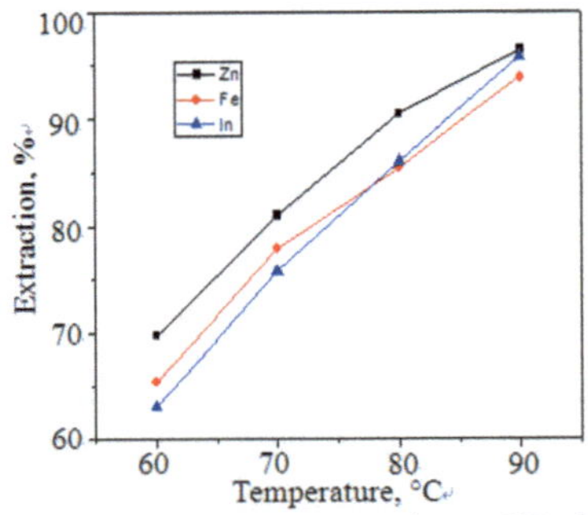

Figure 8. Effect of temperature on extractions of Zn, Fe and In.
(The addition of sphalerite 0.9, 150 g/L H_2SO_4, particle size: 74-58 μm and 5 h).

3.6 Effect of particle size

The effect of particle size on the extractions of Zn, Fe and In were shown in Figure 9. It was showed from Figure 9 that the extractions percentage of Zn, Fe and In increased significantly as the size of particles decreased. This effect was due to the increased in the interfacial area of leaching reaction as the solid particles became smaller. In the following experiments, the particle size of the sample chosen was 74-58 μm to obtain a higher extractions of metal.

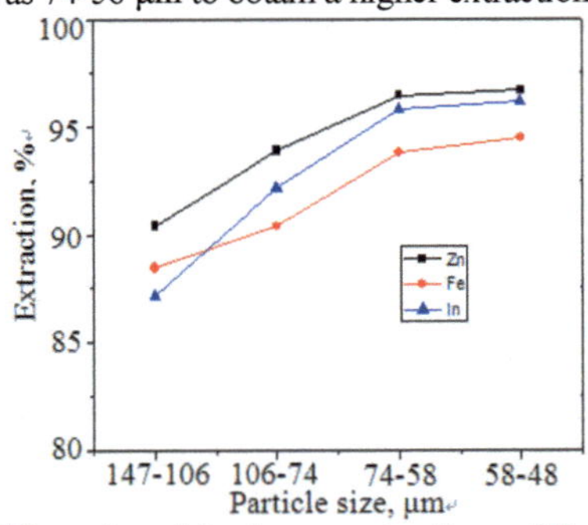

Figure 9. Effect of particle size on extractions of Zn, Fe and In.
(The addition of sphalerite 0.9, 90°C, 150 g/L H_2SO_4 and 5 h).

3.7 Characterization of the simultaneous leaching residue

Simultaneous leach residue, obtained under conditions of addition amount of high-iron sphalerite 0.9, 150 g/L H_2SO_4, particle size 74-58 μm, at 90°C for 240 min, and the data indicated that under the conditions employed up to 96% Zn, 92% Fe and 94% In extractions were achieved. The simultaneous leaching residue was chemically analyzed, and the results were presented in Table II. It can be seen in Table II that the simultaneous leaching residue became an Ag and S^0 rich material with high SiO_2. There were small amounts of Zn, Fe and In.
A typical XRD pattern of the simultaneous leaching residue was given in Figure 10. It can be seen clearly in Figure 10 that constituent phases of the leaching residue changed a lot as compared to the zinc residue and the high-iron sphalerite. The main compositions of the simultaneous leaching residue were elemental Sulphur (S^0) which was the oxidized product of sulphide sulphur, anglesite ($PbSO_4$), sphalerite ((Zn,Fe)S) and quartz (SiO_2). The zinc ferrite was almost entirely dissoluted. There is a small amount of sphalerite in the leaching residue. This may be caused by formation of a layer of elemental sulphur on the surface of concentrate particles which made the dissolve of sphalerite in sulphuric acid solution unavailable.

Table II. The main chemical component of simultaneous acid leaching residue (wt.%)

Zn	Fe	In	S	SiO_2	Ag	Cu	Co	Pb	Sn	CaO	MgO
5.21	5.76	180.6g/t	33.6	19.34	510.9g/t	0.37	<0.005	0.63	0.85	1.07	1.79
Ni	**Cd**	**Al_2O_3**	**Mn**	**As**	**Sb**	**F**	**Cl**	**Ge**	**Ga**	**SO_4^{2-}**	**S^0**
<0.005	0.052	1.63	0.13	1.14	0.0083	0.1	0.01	trace	trace	14.77	28.2

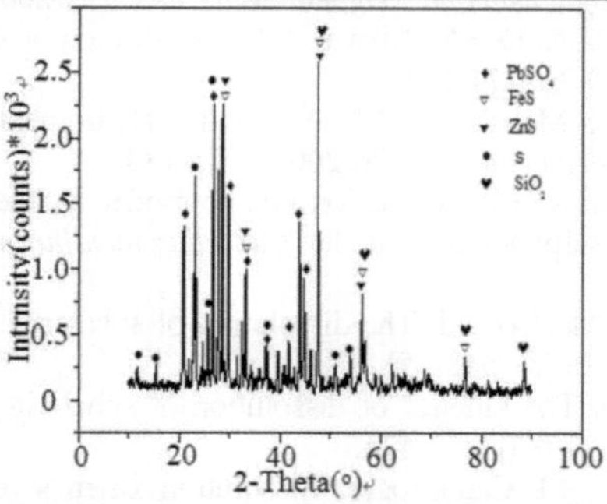

Figure 10. XRD pattern of simultaneous leaching residue.

4. Conclusions

Experiments data indicated that under the conditions with the addition of high-iron sphalerite of 0.9, sulphuric acid concentration of 150 g/L, particle size of 74-58 μm, at leaching temperature of 90°C for 240 min up to 96% Zn, 92% Fe and 94% In extraction were achieved.Under optimal conditions, the main minerals of the simultaneous leaching residue were sulphide sulphur, anglesite, sphalerite and quartz. Zinc ferrite was almost entirely leached.The simultaneous leaching of zinc ferrite in the presence of high-iron sphalerit as reductant was a viable process that effectively extracts zinc and indium and converted Fe^{3+} into Fe^{2+} in solution. However, it was negative influence on the extraction of metals when the high-iron sphalerite was excess. The key point to the simultaneous leaching was the dissolution of sphalerite.

Acknowledgments

This work was supported financially by the National Natural Science Foundation of China (No. 51564030,51474117).

References

1. ISMAEL M.R.C, CARVALHO J.M.R. Iron recovery from sulphate leaches liquors in zinc hydrometallurgy. *Minerals Engineering*, 16 (2003), 31-39.
2. A.M. Alfantazi, R.R. Moskalyk. Processing of indium: a review. *Minerals Engineering*, 16(2003), 687-694.
3. Xuelan Wu, Shunke Wu, Wenqing Qin, Xihong Ma, Yinjian Niu, Shaoshi Lai, Congren Yang, Fen Jiao, Liuyi Ren. Reductive leaching of gallium from zinc residue. *Hydrometallurgy*, 113-114(2012), 195-199.
4. LIANG Duo-qian, WANG Ji-kun, ZHOU Ting-xi. Simultaneous Pressure Oxidation Leaching of Zinc Sulfide Concentrate and Neutral Leaching Residue. *Nonferrous Metals*, 61(4) (2009), 72-76.
5. S.E. James, J.L. Watson, J. Peter. Zinc Production-A Survey of Existing Smelters and Refineries. *Lead-Zinc 2000*, (Warrendale, PA: TMS, 2000), 205-225.
6. B. Parvatheeswara Rao, K.H. Rao. Distribution of In^{3+} ions in indium-substituted Ni–Zn–Ti ferrites. *Journal of Magnetism and Magnetic Materials*, 292(2005), 44-48.
7. Ramachandra Sarma V N, Deo K, Biswas A K. Dissolution of zinc ferrite samples in acids. *Hydrometallurgy*, 2(1976), 171-184.
8. Nathalie Leclerc, Eric Meux, Jean-Marie Lecuire. Hydrometallurgical extraction of zinc from zinc ferrites. *Hydrometallurgy*, 70(2003), 175-183.
9. Filippou D, Demopoulos G P. A reaction kinetic model for the leaching of industrial zinc ferrite particulates in sulphuric acid media. *Canadian metallurgical quarterly*, 31(1) (1992), 41-54.
10. J.E. Dutrizac, R.J.C. MacDonald. The dissolution of sphalerite in ferric chloride solutions. *Metall. Trans.*, 9 (B) (1978), 543-551.
11. G.E. Bobeck, H. Sue. The kinetics of dissolution of sphalerite in ferric chloride solution, *Metall. Trans.*, 16 (B) (1985), 413-424.
12. S. Aydogan, A. Aras, M. Canbazoglu. Dissolution kinetics of sphalerite in acidic ferric chloride leaching. *Chemical Engineering Journal*, 114(2005), 67-72.
13. Heidi Markus, Sigmund Fugleberg, Daniel Valtakari, Tapio Salmia, Dmitry Yu. Murzina, Marko Lahtinen. Reduction of ferric to ferrous with sphalerite concentrate kinetic modeling. *Hydrometallurgy*, 73(2004), 269-282.
14. Cunxiong Li, Chang Wei, Hong-sheng Xu, Minting Li, Xingbin Li, Zhigan Deng, Gang Fan. Oxidative pressure leaching of sphalerite concentrate with high indium and iron content in sulfuric acid medium. *Hydrometallurgy*, 102(2010), 91-94.
15. M.C.B. Fortes, J.S. Benedetto. Technical note sepatation of indium and iron by solvent extraction. *Minerals Engineering*, 11(1998), 447-451.
16. Li Shiqing, Tang Motang, He Jing, Yang Shenghai, Tang Chaobo, Cheng yongming. Extraction of indium from indium-zinc concentrate. *Transactions of Nonferrous Metals Society of China*, 16(2006), 1448-1454.

Rare Metal Technology 2016

Rare Earth Elements / Base & Rare Metals II

Rare Metal Technology 2016
Edited by: Shafiq Alam, Hojong Kim, Neale R. Neelameggham, Takanari Ouchi, and Harald Oosterhof
TMS (The Minerals, Metals & Materials Society), 2016

RECOVERY OF YTTRIUM AND NEODYMIUM FROM COPPER PREGNANT LEACH SOLUTIONS BY SOLVENT EXTRACTION

Rebecca Grace Copp and J. Brent Hiskey

Department of Mining and Geological Engineering
University of Arizona; 1235 E. James E. Rogers Way; Tucson, AZ, 85721, USA

Keywords: Yttrium, Neodymium, Solvent Extraction, Copper PLS

Abstract

The solvent extraction of yttrium and neodymium from copper pregnant leach solutions (PLS) using Primene JM-T, a primary aliphatic amine, has been studied. Effect of contact time, pH, sulfate concentration, and extractant concentration were investigated using synthetic and actual PLS systems. Standard experimental conditions were 5 minute contact time, pH ~2.5, and 10 % v/v Primene JM-T concentration. Distribution isotherms were constructed for the pure systems and for actual copper leach solutions. Synthetic solutions contained 100 ppm Y and ~75 ppm Nd. Samples of PLS contained 2 - 10 ppm Nd and 14 - 30 ppm Y. Results showed that complete extraction of both yttrium and neodymium occurred within five minutes and at pH values greater than 1. It was also found that sulfate concentration does not inhibit extraction at any concentration. Additionally, the distribution isotherms created show that extraction for these metals can operationally take place in one stage.

Introduction

Mine production of rare-earths in the United States currently only comes from one company: Molycorp [1]. Molycorp's Mountain Pass rare-earth processing facility in California processes rare-earths from its bastnasite ore. Bastnasite is a rare-earth fluorocarbonate mineral which can contain yttrium and neodymium. At Mountain Pass, approximately 0.12% of the rare earths found in the bastnasite ore are yttrium. However, in recent years there has been no mine production of yttrium or neodymium in the United States, yet in 2013 and 2014 the United States consumed approximately 200 tons of yttrium oxide each year. This results in a 100% import reliance of this metal. The majority of which is imported from China. This comes at a cost since 99.999% pure yttrium oxide cost $15-17 per kilogram in 2014, and 99.9% pure yttrium metal cost $55-65 in the same year. Concurrently, neodymium oxide with greater than 99% purity was valued at $56-60 per kilogram in 2014.

Both yttrium and neodymium are used in the production of important materials. The main end use of yttrium is as an additive in aluminum and magnesium alloys in order to increase their strength [2]. Other uses for yttrium or yttrium oxide (Y_2O_3) include: as a component in microwave filters for radar, as a component in lasers that are strong enough to cut through metals, in white LED lights, as an additive to glass to make it more resistant to heat and shock, and also in the manufacture of superconductors. The main end use of neodymium is in an alloy with iron and boron in order to produce permanent magnets [3]. Neodymium can also be found in glass used to make lasers and as a catalyst in polymerization reactions. Yttrium and neodymium are both necessary in that they are unique, and so it is very difficult if not impossible to find substitutes for them in the applications that they are used for.

Since both yttrium and neodymium are found in the earth's crust at the low concentration of 0.3 ppm, it is advantageous to extract them whenever present. Additionally, less than 10 percent of both yttrium and neodymium are recycled, so new sources are continually needed. Yttrium, neodymium, and a number of other rare earth elements can be found in pregnant leach solutions (PLS) from copper mining operations. The recovery of yttrium and neodymium specifically from these leach solutions by solvent extraction has been studied.

Desouky and co-workers studied the liquid-liquid extraction of yttrium using Primene JM-T from acidic sulfate solutions [4]. Desouky et. al studied the effects of extractant concentration, diluent type, pH, contact time, and temperature on yttrium extraction along with constructing an extraction isotherm. The extraction was carried out by solvent extraction using a synthetic solution containing 8.86 x 10^{-3} M yttrium. This stock solution was made using yttrium oxide. Baseline conditions for this investigation included a phase ratio of 1:1, room temperature, 5 minute equilibrium time, pH 1.5, and 0.4 M Primene JM-T concentration in the kerosene as the organic phase.

Several conclusions were drawn from the research of Desouky and co-workers. First, aliphatic kerosene proved to be a successful diluent when using Primene JM-T as the extractant. Additionally, increasing temperature had a negative effect on the yttrium extraction as the reaction is exothermic. The optimum number of stages for the yttrium extraction was found to be three. They also studied the stripping of the yttrium off of the organic and determined that optimum stripping occurs using 3 M nitric acid at a phase ratio of 1:1 with a 3 minute contact time.

Additional work has been done to study extractants and their effectiveness on the extraction of rare earth elements. Rice and Stone (1962) investigated the use of primary amines (Primene 81-R and Primene JM-T) in the extraction of rare earth elements [5]. Additionally, Levenson and co-workers (1992) demonstrated the effectiveness of Primene JM-T in the extraction of yttrium at low pH values [6]. It should be noted that when using amines for yttrium extraction, anions are extracted not cations. In the aqueous feed for solvent extraction tests the yttrium, Y^{3+} cation is complexed with the sulfate. The yttrium sulfate anionic complex species are either $Y(SO_4)_2^-$ or $Y(SO_4)_3^{3-}$. Neodymium forms analogous sable anionic complexes with sulfate.

This paper describes the investigation of extracting both yttrium and neodymium from both synthetic solutions and copper mine pregnant leach solutions. This research utilized the hydrometallurgical process of solvent extraction.

Experimental Methods

Materials

Synthetic yttrium and neodymium solutions as well as actual copper PLS from operating copper mines in Arizona were used for this study. A 100 ppm pure yttrium solution was made using yttrium sulfate hydrate, $Y_2(SO_4)_3{\cdot}8H_2O$, and deionized water in order to study the extraction of yttrium from a pure system. Likewise, a ~75 ppm pure neodymium solution was made using neodymium sulfate hydrate, $Nd_2(SO_4)_3{\cdot}11H_2O$, and deionized water for the same purpose with neodymium.

Actual leach solutions from two operating mines in Arizona were collected and analyzed for this study. These solutions were designated PLS *A* and PLS *B*. The composition of these solutions are reported in Table I.

Table I Typical rare earth concentrations in copper leach solutions

PLS	Concentration (ppm)					
	Ce	La	Nd	Pr	Sm	Y
A	1	<1	2	<1	1	14
B	8	1	10	3	4	30

The organic used for all of the experiments consisted of kerosene from Chevron Phillips Mining Chemicals as the diluent and Primene JM-T from Dow Chemical as the extractant. Primene JM-T is an aliphatic, primary amine which is commonly used for extraction of rare earths. For most of the tests the organic was protonated by contacting it with a 1 M sulfuric acid solution before it was used in the solvent extraction tests.

Shake-Out Experiments

All experiments were carried out in 250 mL separatory funnels and shaken using a Wrist Action Shaker. Standard experimental conditions for all of the tests were 5 minute contact time, pH ~2.5, 10 % v/v Primene JM-T concentration, room temperature, and 1:1 organic to aqueous phase ratio. One to one organic to aqueous phase ratios involved using 100 mL of organic and 100 mL of aqueous.

The first step in the experimental process was to make the organic and aqueous solutions. Standard 10% organic was made by mixing 270 mL of kerosene and 30 mL of Primene JM-T. The molarity of a 10% v/v Primene JM-T solution is 0.29 M Primene JM-T. The organic was then protonated by contacting it with 1M sulfuric acid.

A synthetic solution containing 100 ppm yttrium and another synthetic solution containing ~75 ppm neodymium were made for the aqueous solutions. The yttrium solution was made by mixing 3.43 grams of $Y_2(SO_4)_3{\cdot}8H_2O$ with 1000 mL of deionized water. This resulted in a 1000 ppm or 1 g/L yttrium solution. 10 mL of the stock yttrium solution were mixed with 90 mL of DI water for each shake-out test. This made a 1.12 x 10^{-3} M yttrium solution. The neodymium solution was made by mixing 2 grams of $Nd_2(SO_4)_3 \cdot 11H_2O$ with 1000 mL of DI water. This resulted in a 1000 ppm or 1 g/L neodymium stock solution from which 10 mL was taken to be mixed with 90 mL of DI water for each shake-out test involving neodymium. These solutions were 6.93 x 10^{-4} M neodymium solutions.

Sample Analysis

After each shake out test, ~30 mL of the resulting aqueous solution (raffinate) was taken from the separatory funnel and sent to SGS Metcon in Tucson, Arizona for analysis. At Metcon, an ICP was used to determine the concentration of the desired metal in the sample. From there the concentration extracted into the organic could be determined using a mass balance knowing the feed concentration and the aqueous raffinate concentration, and then Equations 1 and 2 could be used to determine the distribution coefficient and the percent extraction, respectively, where M is the desired metal.

$$D = \frac{[M]_{org}}{[M]_{aq}} \tag{1}$$

$$\% \, Ext = \frac{\frac{O}{A} D}{1 + \frac{O}{A} D} \tag{2}$$

Results and Discussion

Effect of Time

The first series of experiments using 100 ppm yttrium were performed to determine the time for the yttrium extraction to reach equilibrium. The standard experimental conditions were used except shake outs were done for a duration of 1, 2, 5, 10, 15, 30, and 60 minutes. These kinetic tests were performed using organic that had not been protonated prior to use. The results indicated that the extraction of yttrium reached equilibrium in about 1 minute. A shake out time of 5 minutes was used as a standard condition in subsequent tests. The effect of protonating the organic prior to each set of tests was evident in these experiments. Without protonation, yttrium extraction was 81% after 5 minutes of contact. On the other hand, with protonation of the organic, yttrium extraction was 99% after 5 minutes. Similar results were observed for neodymium.

Effect of pH

In the second set of experiments with the synthetic yttrium and neodymium solutions, the effect of the pH of the aqueous solution on the extraction was studied by using the standard experimental conditions but varying the pH from 0.5 to 3.5. The results of these tests are depicted in Figure 1. Extraction is very poor at low pH values. At pH 0.5 the extraction of yttrium and neodymium was only 30% and 50%, respectively. At pH 1 and higher up to 3.5 the extraction ranged from 97% to 99%. At pH values greater than 1 and up to 3.5 complete extraction occurs and is independent of pH. Results of Levenson et al. (1992) using 10^{-3} M Y, 0.5 M Na_2SO_4, and 5% Primene JM-T in Escaid diluent were nearly identical to those shown here.

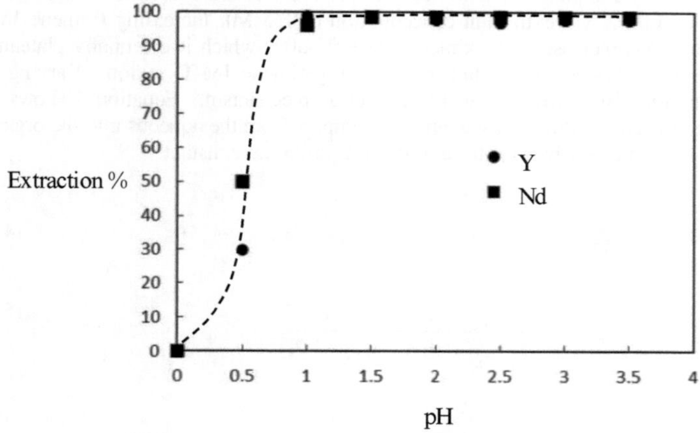

Figure 1 Effect of pH using synthetic yttrium and neodymium solutions

Effect of Sulfate Concentration

Actual copper leach solutions can vary widely with respect to total sulfate concentration. Sulfate concentrations ranging from 0.3 M to 1.3 M have been measured for leaching operations in the Southwestern United States. The effect of sulfate concentration in the aqueous phase on the extraction was studied using the synthetic yttrium solution. The sulfate concentration was adjusted by adding sodium sulfate, Na_2SO_4, to the aqueous phase. The sulfate concentrations ranged from 0.01 – 1.0 M SO_4^{2-}. The results of these experiments are provided in Table II. As can be seen, the extraction of yttrium is practically complete at all sulfate concentrations tested and can be considered essentially independent of sulfate concentration. Copper PLS can be high in sulfate concentration without adversely affecting the extraction of yttrium.

Table II Effect of sulfate concentration using synthetic yttrium solution

$[SO_4^{2-}]$ (M)	Raffinate [Y] (ppm)	% Extraction
0.01	3	97
0.05	2	98
0.1	7	93
0.5	1	99
1.0	4	96

Effect of Extractant Concentration

The effect of Primene JM-T concentration in the organic phase was studied in the range of 0.015 - 0.59 M (0.5 to 20 v/v %). Figure 2 shows percent extraction versus Primene JM-T concentration. Practically complete extraction occurred at every concentration level. The lowest recover was 90% at a 0.5% v/v extractant concentration (0.015 M). Increasing Primene JM-T concentration up to 2.5% increased the extraction to 99% after which it essentially plateaued. The insert in Figure 2 provides more detail for the low [Primene JM-T] region. Varying the extractant concentration also helps explain the extraction mechanism. Equation 4 shows the reaction where the yttrium is extracted as an anionic complex from the aqueous into the organic. Equations 5 and 6 show relationships for the extraction equilibrium constant.

$$(RNH_3)_2 \cdot\cdot SO_4 + 2Y(SO_4)_2^- = 2RNH_3 \cdot\cdot Y(SO_4)_2 + SO_4^{2-} \quad (4)$$

$$K_{eq} = \frac{[RNH_3 \cdot\cdot Y(SO_4)_2]^2\ [SO_4^{2-}]}{[Y(SO_4)_2^-]^2\ [(RNH_3)_2 \cdot\cdot SO_4]} \quad (5)$$

$$\log K_{eq} = 2\log D + \log[SO_4^{2-}] - \log[(RNH_3)_2 \cdot\cdot SO_4] \quad (6)$$

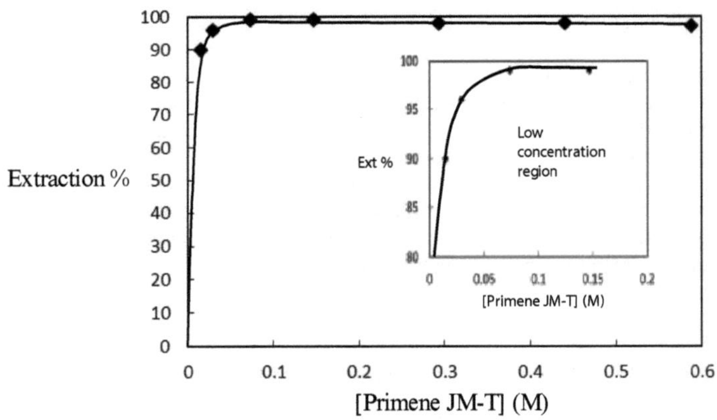

Figure 2 Effect of extractant concentration using synthetic yttrium solution

Plotting the distribution coefficient (D) versus the Primene JM-T concentration on a log-log plot is shown in Figure 3 for the pre-plateau region. This plot indicates a slope of 1.5 which is close to the two moles of Primene JM-T required for each mole of Y^{3+} extracted. Additionally, at 0.5% v/v extractant concentration, the organic can hold approximately 1300 ppm of yttrium.

Therefore, the organic never reaches anywhere near its full capacity. Neodymium extraction was complete (~99%) for all Primene JM-T concentrations between 0.015 and 0.59 M.

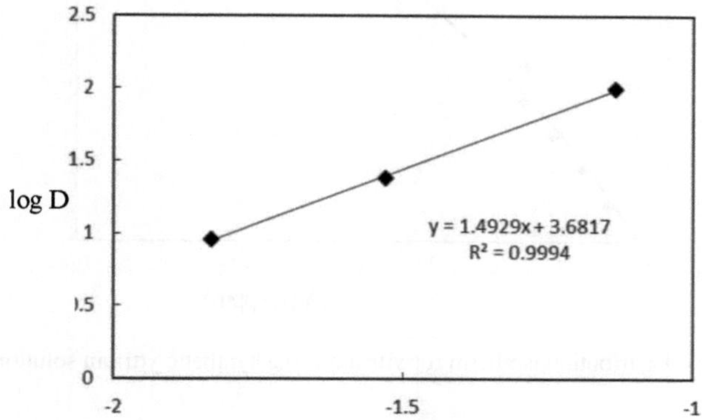

Figure 3 Distribution coefficient versus Primene JM-T concentration using synthetic yttrium solution.

Distribution Isotherm

Finally, a distribution isotherm for the extraction of yttrium using the synthetic yttrium solution was generated. The isotherm was created by using the standard experimental conditions and varying the organic to aqueous phase ratio. O/A phase ratios tested included 1:1, 1:2, 1:3, 1:4, 1:5, and 1:10. The isotherm can be seen in Figure 4. At the normal 1:1 O/A phase ratio, 99% of the yttrium was extracted. As the aqueous portion increased from 1 to 10 the recovery slowly decreased from 99% to 89%. The isotherm shows that the extraction of yttrium at these conditions can operationally take place in one stage.

A distribution isotherm for synthetic solutions of neodymium was also generated. Under the same conditions, the results for neodymium were similar to those for yttrium.

The isotherm shown in Figure 4 is replotted along with the isotherm generated by Desouky et al in Figure 5.

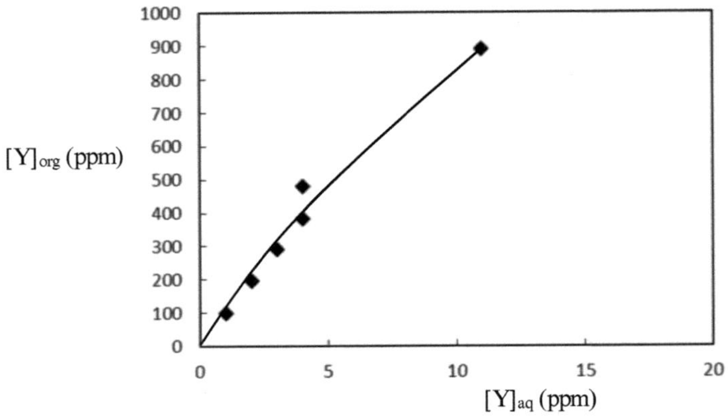

Figure 4 Distribution isotherm for yttrium using synthetic yttrium solution

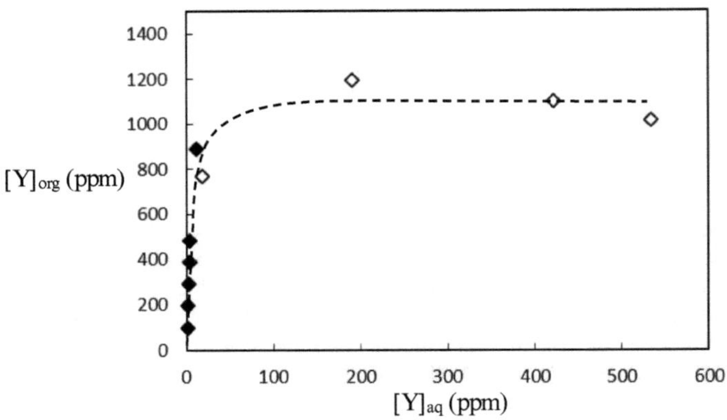

Figure 5 Distribution isotherm for low and high Y concentrations

In Figure 5 the open diamond markers show the isotherm that Desouky et.al. generated, while the closed diamond markers show the isotherm produced from this work. The isotherm created in this research shows the low Y concentration behavior. The concentration of yttrium used in this research was approximately 8 times lower than the concentration used by Desouky and co-workers. Their research also showed that the percent extraction does decrease as the aqueous volume increases, with recovery decreasing from 98% to 66% as the ratio changed from 1:1 to 1:4.

Throughout all of the tests with the synthetic yttrium solution, it was found that while using the standard experimental conditions practically all of the yttrium was recovered in each test. Because of this, it was decided to test multiple contacts of the organic with fresh aqueous feed each time. For this test an organic to aqueous ratio of 1:2 was used along with 5% v/v extractant concentration in the organic. Standard conditions of pH ~2.5, 5 minute contact time, and room temperature were still maintained. For this test, the organic was contacted with fresh aqueous feed four times. The isotherm created is shown in Figure 6 with the open square markers and the dashed line. The organic was successfully able to continue to extract yttrium up to four contacts with fresh feed.

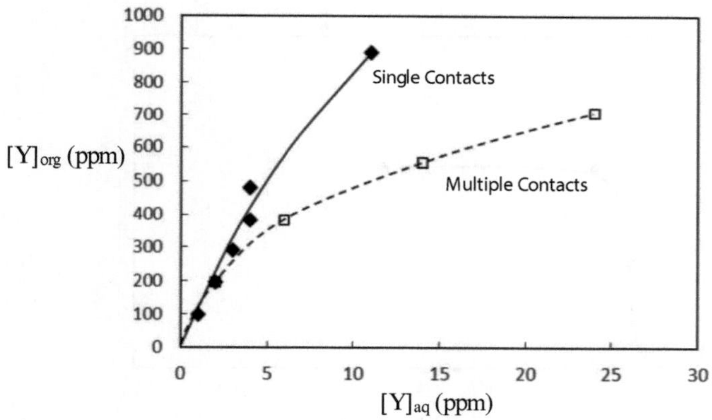

Figure 6 Comparison of distribution isotherms for yttrium using single and multiple contacts

Extraction of Yttrium and Neodymium from Copper PLS

After studying the extraction of these metals from synthetic solutions, extraction from actual copper PLS was studied. PLS *A* was tested at a pH of 1.32 using standard test conditions. Table III provides test results for PLS *A* extraction with Primene JM-T at different organic to aqueous ratios.

Table III Extraction of yttrium and neodymium from PLS *A*

PLS	pH	O/A	Yttrium			Neodymium		
			Feed (ppm)	Raffinate (ppm)	% Ext	Feed (ppm)	Raffinate* (ppm)	% Ext
A	1.32	4:1	14	1	93	2	< 1	75
		2:1	14	1	93	2	< 1	75
		1:1	14	2	86	2	< 1	75
		1:2	14	8	46	2	1	50
		1:4	14	14	0	2	2	0

*Raffinate values less than 1 ppm are assumed to be ~ 0.5 ppm.

Extraction was achieved at all O/A ratios except 1:4 for both yttrium and neodymium. The yttrium isotherm for the data showing some degree of extraction is shown in Figure 7. It appears that under these conditions the organic reaches maximum loading with respect to yttrium at approximately 12 ppm [Y].

Three experiments using PLS *B* were conducted at standard conditions. These results indicated no extraction of either yttrium or neodymium. This is explained by the low pH (1.16) of the PLS and the extremely high concentrations of Al and Fe which have the potential of forming anionic complexes with sulfate and thus hindering the extraction of the low concentration REEs.

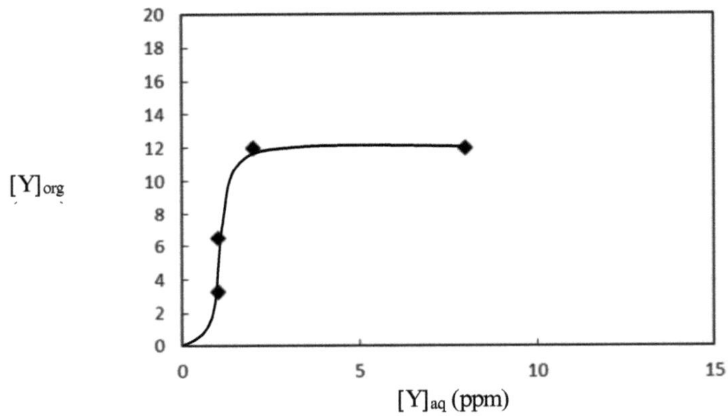

Figure 7 Yttrium distribution isotherm for PLS A

Conclusions

Recovery of yttrium and neodymium with Primene JM-T was successfully demonstrated for synthetic solutions. It was found that protonating the organic before contacting the organic with aqueous to extract the desired metal was critical. It was necessary for the organic to take a proton from the sulfuric acid before it entered into a reaction with the yttrium or neodymium. The extraction of yttrium and the extraction of neodymium proved to be very similar under standard conditions. The extraction of both yttrium and neodymium happens very quickly. Extraction of these metals is very poor at pH < 1, but extraction is nearly complete and independent of pH in the range of 1 to 3.5. PLS usually has a pH of 2-3, so pH will not hinder extraction from leach solutions. Additionally, PLS can have high sulfate, but the extraction of these metals is completely independent of sulfate concentration. Results for actual PLS solutions were found to be particularly sensitive to the overall solution chemistry of the sample.

References

1. Gambogi, Joseph. "USGS Mineral Commodity Summary 2015." U.S. Department of the Interior, 2015.

2. Emsley, John. "Yttrium." *Periodic Table*. Royal Society of Chemistry, 2014.

3. Emsley, John. "Neodymium." *Periodic Table*. Royal Society of Chemistry, 2014.

4. Desouky, O.A., A.M. Daher, Y.K. Abdel-Monem, and A.A. Galhoum. "Liquid-Liquid Extraction of Yttrium Using Primene-JMT from Acidic Sulfate Solutions." *Hydrometallurgy* 96 (2008): 313-17.

5. Rice, A.C., and C.A. Stone. "Amines in Liquid-Liquid Extraction of Rare-Earth Elements." (1962).

6. Levenson, Yvonne. "Yttrium Behavior in Amine Solvent Extraction." (1992).

Rare Metal Technology 2016
Edited by: Shafiq Alam, Hojong Kim, Neale R. Neelameggham, Takanari Ouchi, and Harald Oosterhof
TMS (The Minerals, Metals & Materials Society), 2016

CALCINED NANOCRYSTALINE LAYERED DOUBLE HYDROXIDES FOR THE REMOVAL OF ARSENATE AND ARSENITE

Eman Wahbah, Yousef Mohassab, Manoranjan Misra, Monalisa Panda

Department of Metallurgical Engineering, University of Utah, Salt Lake City, Utah 84112, USA

Keywords: LDH, Arsenic, Adsorption, Rare earth elements

Abstract

This work studies the removal of arsenic (V), arsenate, and arsenic (III), arsenite, from aqueous solution using calcined Quintinite-3T. The two adsorbents Quintinite-3T was prepared, calcined at different temperatures, and characterized by XRD, BET, and SEM. The surface area of the calcined Quintinite-3T was 193.4 m^2/g. The particles size ranged from 12 to 56 nm for the uncalcined Quintinite-3T and from 24 to 42 nm for the calcined Quintinite-3T. Kinetic analysis, adsorption isotherm, and factors affecting the adsorption were investigated. Calcined Quintinite-3T retained As(v) and did not release it back to water even after an entire month.

Introduction

Arsenic occurs in the drinking water of many countries all over the world, such as Bangladesh, West Bengal, Argentina, Vietnam, China, Mexico, Nepal, India, Hungary, Australia, USA, and Canada, in the form of As(III), arsenite, and As(V), arsenate [1-4]. As(III) is more difficult to be removed than As(V), but fortunately As(III) exists mostly in anaerobic ground water [5]. The occurrence of arsenic in the ground water is mainly due to minerals dissolved naturally from weathered rocks and soils over time or by human activities [6]. Arsenic is very harmful for humans, animals, and plants. Long-term exposure to As can cause hyperkeratosis, skin lesions, and various types of cancers [7], including skin and bladder cancers [8]. The Environmental Protection Agency (EPA) of the US limited arsenic concentration in drinking water to 50 ppb, and they changed this limit later to 5 ppb [8]. Many researchers have studied arsenic removal by various adsorbents, such as steel making slag [1], mesoporous alumina [9], zirconium-manganese binary hydrous oxide [10], rutile ore [11], layered double hydroxides (LDHs) [7, 12], and some nanofibers as titanate nanofibers [13], titania nanofibers [14], and polyvinyl alcohol/Fe^{3+} nanofibers [5].
In this study calcined Quintinite-3T adsorbent was investigated for the removal of As(V) and As(III) from water. Q-3T was calcined at different temperatures: 200 °C, 300 °C, 400 °C, and 500 °C. The effect of aging on calcined Q-3T was explored from 30 min to 1 month. The adsorbent was characterized before and after adsorption. In addition, new adsorbent (calcined mixture of Q-3T and diatomaceous powder) was tested for arsenic removal. Samples were characterized by SEM, XRD, and BET surface area measurement. Arsenic was quantitatively determined by ICP-MS [15].

Experimental Work

All chemicals were used without further purification. Arsenate solution was prepared from $Na_2HAsO_4.7H_2O$ (98%) and deionized water (18.2 M cm). Arsenite solution was prepared from As_2O_3 (99.996%) and deionized water.

Adsorbent Preparation:

The Al/Mg solution was prepared by dissolving 56.98 g of $Mg(NO_3)_2.6H_2O$ and 41.68 g of $Al(NO_3)_3.9H_2O$ in 100 mL of deionized water. The NaOH solution was prepared by dissolving 20.00 g of NaOH to 13.626 g of Na_2CO_3 in 100 mL of deionized water. Then 45 mL of Mg/Al solution were added to 70 mL of NaOH solution under stirring at 400 rpm and temperature of 60 °C for 24 hours in an oil bath. The precipitate was washed by deionized water, dried at 120 °C, and then calcined at 500 °C. Samples were characterized by SEM, XRD, and BET surface area measurement. Arsenic was quantitatively determined by ICP-MS. A weight of 0.416 g of $Na_2HAsO_4.7H_2O$ and 0.132 g of As_2O_3 were dissolved each in a 100 mL DI water to prepare 1000 ppm of As(V) and As(III), respectively. Then this solution was diluted carefully with thorough mixing to obtain a 100 ppb of arsenic solution. Further details of the experiment can be found elsewhere [15].

Results

Adsorbent Characterization:

The as-synthesized Q-3T (uncalcined) showed peaks at 2Θ of 11.64°, 23.37°, 34.42°, 38.55°, 45.71°, 60.72°, and 62.06°, as shown in Fig. 1(a). These peaks matched uncalcined Q-3T structure as reported in the literature [4, 7, 17, 18]. The calcined Q-3T at 200 °C showed different peaks than the uncalcined Q-3T.The peaks at 2Θ of 11.64°and 23.37° in uncalcined Q-3T became broader and shifted to 13.21° and 26.57°, respectively, in the calcined Q-3T, whereas peaks at 2Θ of 34.42°, 38.55°, and 45.71° were combined in one broad peak at 35.59°. Also, the peaks at 2Θ of 60.72°and 62.06° turned into one broad peak at 60.43° in the calcined Q-3T, as Fig. 1(b) shows. When Q-3T was calcined at 300 °C, there was no difference in XRD peaks from the calcined Q-3T at 200 °C except that peaks are broader at the higher temperature, as presented in Fig. 1(c). At 400 °C and 500 °C of calcination, shown in Fig. 1(d and e), respectively, broad peak were observed at 2Θ of 35.27°, corresponding to amorphous alumina. In addition, two additional peaks at 2Θ of 42.92° and 62.50° existed that corresponding to MgO (periclase). Therefore, by calcination at temperatures higher than 400 °C, Q-3T should be converted to Al and Mg mixed oxides.

After the adsorption of As(V) by calcined Q-3T, its layers were regenerated, and the adsorbent retrieved its original configuration before calcination as Fig. 2(a) shows. This is attributed to the rehydration during immersing in the aqueous arsenic solution, which is called "memory effect." This is evidenced by the similarity of the XRD patterns of the adsorbent after 30 min of immersion in the aqueous solution and those of the uncalcined Q-3T except for one peak at 2Θ of 11.64° that was not observed after the 30 min rehydration, as presented in Fig. 2(b). This could be attributed to the fact that 30 min immersion was too short to completely regenerate the Q-3T. On the other hand, immersion of the adsorbent in the aqueous solution for a week and up to a month helped the adsorbent regain its original structure before calcination as Fig. 2 shows. To further verify this observation, the adsorbent was tested after immersion in the As(III) aqueous solution and analyzed by XRD at the same time intervals. It was found that the adsorbent retrieved its original crystal structure after a week of the hydration and up to three weeks. However, after 1 month of immersion the peak at 2Θ of 11.64° disappeared.

Specific Surface Area (BET):

The specific surface area of uncalcined Q-3T, calcined Q-3T at different temperatures (200, 300, 400, and 500 °C), Q-3T after the adsorption process was determined by BET. The uncalcined Q-3T has the smallest surface area (64.1 m^2/g), and after calcination the surface area increases with increasing the temperature. The optimum temperature is 500 °C at which the Q-3T has the largest surface area (193.4 m^2/g). After adsorption of As(V) and As(III) by the calcined Q-3T, its surface area decreases due to the expected filing by the adsorbate.

Scanning Electron Microscope (SEM) and Energy Dispersive Spectroscopy (EDS):
The morphology of Q-3T adsorbent before and after calcination and after adsorption was examined by SEM. Figure 3(a) shows the morphology of as-synthesized Q-3T as clusters of particles with

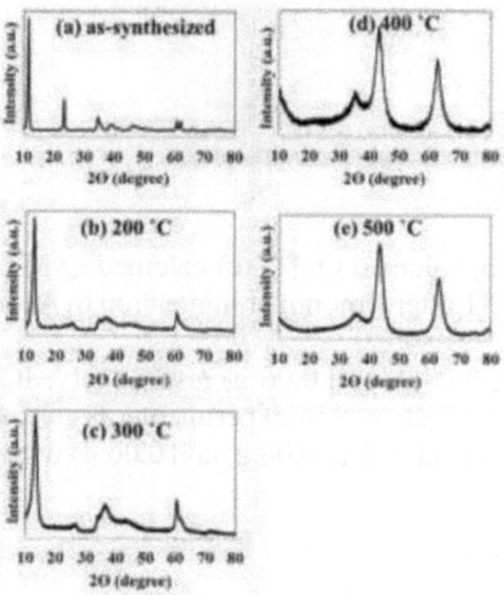

Fig. 1. XRD patterns of (a) as-synthesized Q-3T (uncalcined), (b) Q-3T calcined at 200 °C, (c) Q-3T calcined at 300 °C, (d) Q-3T calcined at 400 °C, and (e) Q-3T calcined at 500 °C.

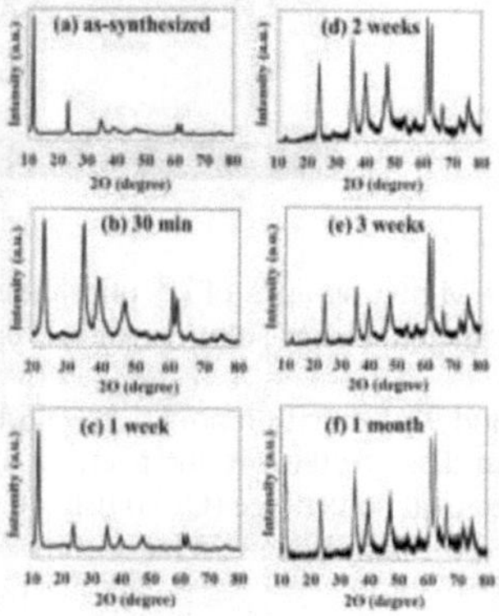

Fig. 2. XRD patterns of (a) as-synthesized Q-3T (uncalcined), (b) calcined Q-3T after 30 min of adsorption of As(V), (c) calcined Q-3T after 1 week of adsorption of As(V), (d) calcined Q-3T after 2 weeks of adsorption of As(V), (e) calcined Q-3T after 3 weeks of adsorption experiment of As(V), and (f) calcined Q-3T after 1 month of adsorption experiment of As(V).

less porosity than calcined Q-3T, shown in Fig. 3(b). Hydrating calcined Q-3T for 30 min and 1 month has altered the morphology of the adsorbent, as seen in Fig. 3(c and d). The platelet-shaped structures were observed clearly after the 1 month immersion, as shown in Fig. 3(d). The particle size as determined from SEM images ranged from 12 to 56 nm for as-synthesized Q-3T and from 24 to 42 nm for the calcined Q-3T. The molar ratio of Mg:Al was found to be 2.0:1.0 in the calcined Q-3T (at 500 °C) as determined by EDS analysis (figure is not shown here).

Adsorption Test:

Adsorption of As(V)

Calcined Q-3T was capable of removing As(V) from aqueous solution after 30 min to 99.8%,

Fig. 3. SEM micrographs of (a) uncalcined Q-3T, (b) calcined Q-3T, (c) calcined Q-3T after 30 min of immersion in As solution, and (d) calcined Q-3T after 1 month of immersion in As solution.

99.7%, and 98.4% with amount of adsorbent 0.1 g, 0.08 g, and 0.06 g, respectively. It is to be noted that these results are the average of a duplicated adsorption experiments. Figure 4 shows As(V) removal percentage (Q_t %) using 0.1 g of adsorbent (0.1 g, 0.08 g, and 0.06 g) with time as calculated by the following relationship:

$$Q_t\% = \frac{(c_0 - c_t)}{c_0} \times 100 \qquad (1)$$

where Q_t % is the As removal percentage at time t and c_0 and c_t are the initial concentration and concentration at time t, respectively. It was found that the As removal increases with increasing the amount of adsorbent and also the adsorbent can conserve its efficiency with time because the increasing or the decreasing in arsenic concentration from 30 min to 1 month within 5 ppb, and this is the allowed concentration limit in the drinking water.

Adsorption of As(III)

Calcined Q-3T removed As(III) from aqueous solution after 30 min by 84.8%, 87.6%, and 95.9 % with the amount of adsorbent 0.1 g, 0.08 g, and 0.06 g, respectively. Figure 5 shows the As(III) removal percentage (Qt %) with different amounts of adsorbent (0.1 g, 0.08 g, and 0.06 g) with time. It is observed that calcined Q-3T removed As(V) better than As(III) because As(III) is more difficult to be removed than As(V) [14]. In contrast to As(V), the removal of As(III) by this adsorbent decreased with increasing the amount of this adsorbent. By increasing the contact time from 30 min to 1 month, As(III) was released back from the adsorbent to the solution (about 35% of the adsorbed As(III) was released back to the solution).

Adsorption of As from contaminated real water

In addition to the synthetic As(V) and As(III) solutions, natural As-contaminated water was examined. The arsenic concentration in this water was determined by ICP-MS to be 93.3 ppb. Two different doses of the calcined Q-3T were used: 0.1 g and 0.08 g. Doses of 0.1 g and 0.08 g of calcined Q-3T could remove about 87.6% and 81.8%, respectively, of the arsenic in this water after 30 min. The arsenic uptake increased with time. The arsenic removal percentage in natural

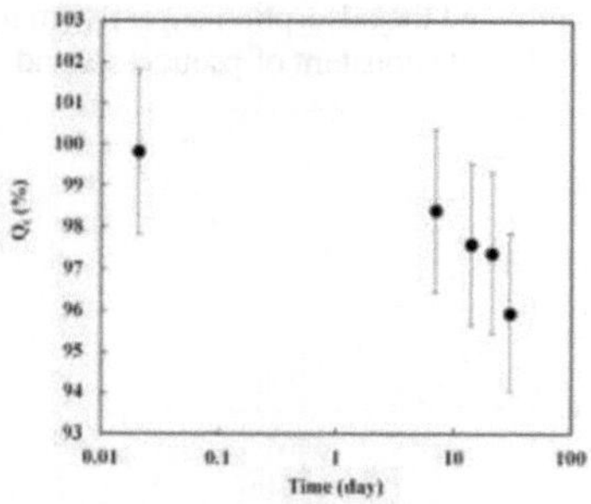

Fig. 4. Removal of As(V) with calcined Q-3T (0.1 g) from aqueous solutions with time. (Error bars represent the % deviation from the average valuesof at least two repeats).

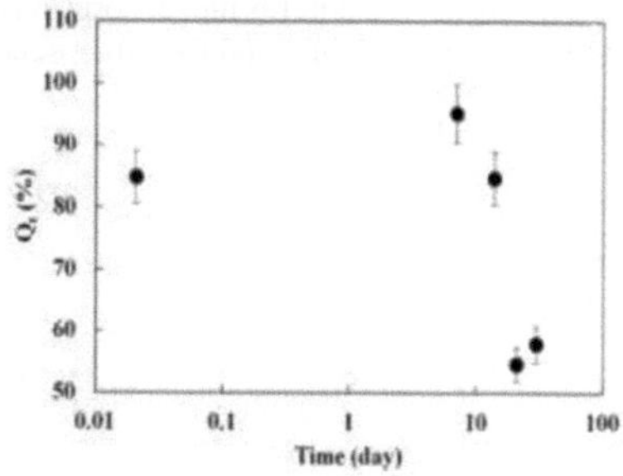

Fig. 5. Removal of As(III) with calcined Q-3T (0.1 g) from aqueous solutions with time.

water was different from that of the synthetic arsenic solution due to the presence of competing anions in the natural water that decreased the arsenic uptake.

Kinetic of As(V) and As(III) Adsorption

The adsorption capacity (q_t) of As(V) and As(III) at time t can be obtained from the following equation:

$$q_t = \frac{(c_0 - c_t)v}{m} \qquad (2)$$

where c_0 and c_t are the initial concentration and the concentration of arsenic solution at time t, respectively, in mg/L; v is the volume of arsenic solution (L); m is the mass of the adsorbent (g); and q_t is the adsorption capacity of arsenic solution at time t (mg/g).

The equilibrium adsorption capacity (q_e) of As(V) and As(III) was determined by plotting the adsorption capacity of arsenic solution at time t (mg/g) versus time t (min), as Fig. 6 shows. The q_e was obtained at 180 min for both As(V) and As(III).

The pseudo-first order, the pseudo-second order, and intraparticle diffusion models were tested against our data to determine the kinetics controlling the adsorption process.

Pseudo-first order model

$$\log(q_e - q_t) = \log q_e - \frac{k_1}{2.303}t \qquad (3)$$

where, q_e and q_t are the equilibrium adsorption capacity (mg/g) and the adsorption capacity (mg/g) at time t (min), respectively, and k_1 is the pseudo-first-order rate constant. The value of k1 can be determined from the slope of the linear plot of log (qe-qt) versus time t [19]. The equilibrium adsorption capacity (q_e) can be determined from the intercept of the same.

Pseudo-second order model

$$\frac{t}{q_t} = \frac{1}{k_2 q_e^2} + \frac{1}{q_e} t \qquad (4)$$

where, q_e and q_t are the equilibrium adsorption capacity (mg/g) and the adsorption capacity (mg/g) at time t (min), respectively, and k_2 is the pseudo-second-order rate constant of pseudo-second

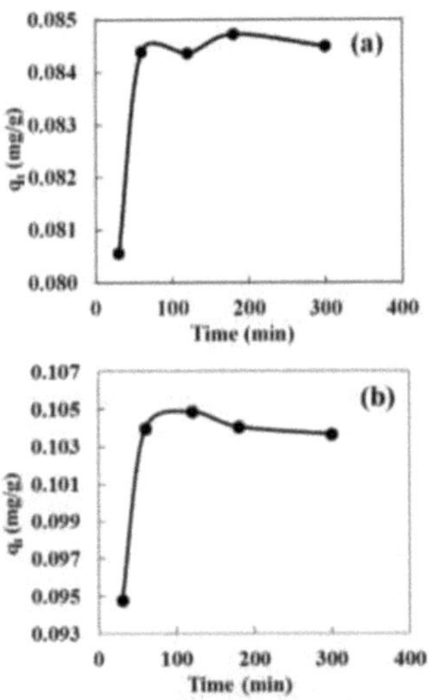

Fig. 6. The equilibrium adsorption capacity (q_e) determination for (a) As(V) and (b) As(III).

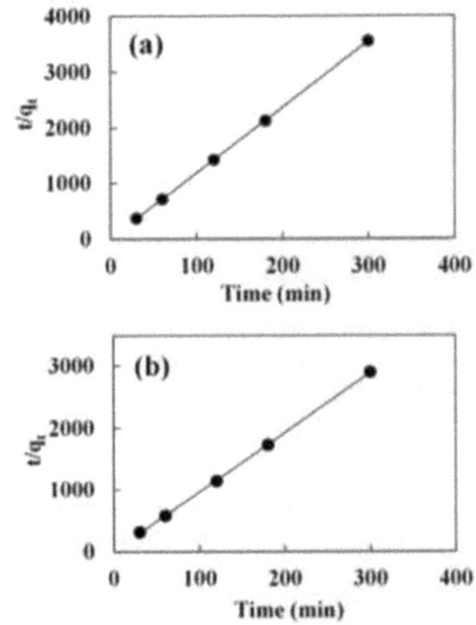

Fig. 7. The pseudo-second order plot of the adsorption of (a) As(V) and (b) As(III) by calcined Q-3T.

order, which can be determined from the intercept of the linear plot of t/q_t versus time t. The equilibrium adsorption capacity (q_e) can be determined from the slope of the same linear plot.

Intraparticle diffusion model

$$q_t = k_3 t^{1/2} + c \qquad (5)$$

where q_t is the adsorption capacity (mg/g) at time t (min). k_3 is the intraparticle-diffusion rate constant and c is also a constant and their value can be respectively determined from the slope and intercept of the linear plot of q_t versus $t^{1/2}$.

By comparing the kinetic parameters of these three models and the correlation coefficients (R^2) for the adsorption of As(V) and As(III) by calcined Q-3T, it was found that the pseudo-second order model best fitted model depending on the correlation coefficients (R^2), as shown in Fig. 7 and the agreement between q_{cal} (calculated) and q_{exp} (experimental) that was observed.

Adsorption Isotherms

In order to determine the adsorption of As(V) and As(III) on calcined Q-3T, the Langmuir and Freundlich models were applied to the data.

Langmuir isotherm model

This isotherm implies that one layer of adsorbate is formed on the surface of adsorbent [18]. This model is formulated as follows:

$$q = \frac{q_{\max} b.c_e}{1 + b.c_e} \quad (6)$$

$$\frac{c_e}{q_e} = \frac{1}{q_{\max}.b} + \frac{c_e}{q_{\max}} \quad (7)$$

where b and q_{max} are the adsorption equilibrium constant and the maximum adsorption capacity, respectively. They can be determined respectively from the intercept and slope of the linear plot of c_e/q_e versus c_e. c_e is the equilibrium adsorption concentration (mg/L), and q_e is the equilibrium adsorption capacity (mg/g).

Freundlich isotherm model

This isotherm model is applicable to multilayer adsorption on the surface of the adsorbent [19], which is expressed as:

$$q_e = K_f.c_e^{\frac{1}{n}} \quad (8)$$

$$\log(q_e) = \log(k_f) + \frac{1}{n}\log(c_e) \quad (9)$$

where k_f and n are the Freundlich coefficients that are defined as adsorption capacity and adsorption intensity, respectively, which can be obtained from the intercept and slope of the linear plot of log q_e against log c_e, respectively. c_e is the equilibrium adsorption concentration (mg/L), and q_e is the equilibrium adsorption capacity (mg/g).

The Langmuir model fitted the data better than the Freundlich model for As(V) and As(III) adsorption on calcined Q-3T, indicating that the adsorbate, either As(V) or As(III), forms a monolayer of on the surface of the calcined Q-3T adsorbent (figures are not shown here).

Factors Affecting As Adsorption

Effect of adsorbent dosage

It was found that 0.1 g of calcined Q-3T offered the highest adsorption of As(V). The order of removal efficiency with respect to adsorbent dose was 0.06 g < 0.08 g < 0.1 g. However, in the case of As(III), 0.06 offered the highest As removal with the exact opposite trend to that of As(V). This is attributed to the difference in dissociation constants.

Effect of aging

As mentioned earlier, calcined Q-3T could conserve its efficiency over time up to a month as evidence by the fluctuation of As concentration in the solution within 5 ppb over the period from 30 min to a month. On the other hand, As(III) concentration increased in the solution with the time as it was released back from the adsorbent to the solution, as discussed earlier.

Effect of pH

The pH of the solution influences the adsorption process significantly. IT was found that the optimal pH for As(V) adsorption by calcined Q-3T was between pH 8 and 10. Beyond this range, the adsorption of As(V) decreases as at higher pH, the OH^- ions increase in the solution; therefore, they compete with $(AsO_3)^{3-}$ ions. At lower pH, H^+ ions increase in the solution that react with nonstoichiometric Al/Mg mixed oxides, leading to form solid metal hydroxide; therefore, atmospheric CO_3^{2-} ions' adsorption are more favorable at the interlayer surface where it sterochemically fits [4]. However, the optimum pH for As(III) adsorption by calcined Q-3T was 10.4.

Effect of calcination

Q-3T was calcined in an electric resistance furnace at temperatures of 200 °C, 300 °C, 400 °C, and 500 °C. It was found that arsenic adsorption increases with the calcination temperature, as presented in Fig. 15. It worth noting that calcination with microwave was attempted, but the XRD patterns showed no evidence of calcination.

Conclusions

Quintinite-3T (Q-3T) was synthesized and characterized by SEM, XRD, and BET. Calcined Q-3T was tested for removing As(V) and As(III) from an aqueous solution. An amount of 0.1 g of this adsorbent was found to remove 99.8 % of 100 ppb of As(V) solution and 84.8 % of 100 ppb of As(III) solution within 30 min. The optimal adsorbent dose of was 0.1 g for As(V) and 0.06 g for As(III). The BET surface area increased by increasing the calcination temperature from 200 °C to 500 °C. As a result, the maximum As adsorption was obtained by the Q-3T calcined at 500 °C. The As adsorption kinetics was controlled by the pseudo-second order model. The Langmuir isotherm better fitted the data than the Freundlich model. The optimal pH for As(V) and As(III) adsorption by calcined Q-3T were 9.7 and 10.4, respectively. The calcined Q-3T regained its original structure after immersion in arsenic aqueous solution due to "memory effect." The adsorbent conserved its efficiency with time until 1 month with As(V), but As(III) was released back from the adsorbent to the solution with time.

References

1. C. Oh, S. Rhee, M. Oh, and J. Park, "Removal Characteristics of As(III) and as(V) from Acidic Aqueous Solution by Steel Making Slag." *J. of Hazard. Mat.*, 213-214 (2012), 147-155.
2. T. Bujdoso, A. Patzko, Z. Galbacs, and I. Dekany, "Structural Characterization of Arsenate Ion Exchanged Mgal-Layered Double Hydroxide." *Appl. Clay Sci.*, 44 (2009), 75-82.
3. X. Wu, X. Tan, S. Yang, T. Wen, H. Guo, X. Wang, and A. Xu, "Coexistence of Adsorption and Coagulation Processes of Both Arsenate and Nom from Contaminated Groundwater by Nanocrystallined Mg/Al Layered Double Hydroxides." *Water Res.*, 47(2013), 4159–4168.
4. M. Chetia, R.L. Goswamee, S. Banerjee, S. Chatterjee, L. Singh, R.B. Srivastava, and H.P. Sarma, "Arsenic Removal from Water Using Calcined Mg-Al Layered Double Hydroxide." *Clean Technol. Environ. Policy*, 14 (2012), 21-27.
5. N. Mahanta and S. Valiyaveettil, "Functionalized Poly (Vinyl Alcohol) Based Nanofibers for the Removal of Arsenic from Water." *RSC Advances*, 3 (2013), 2776-2783.

6. E.A.M. Abdallah and G.A. Gagnon, "Arsenic Removal from Groundwater through Iron Oxyhydroxide Coated Waste Products." *Can. J. Civ. Eng.*, 36 (2009), 881-888.
7. K.H. Goh, T.T. Lim, and Z.L. Dong, "Removal of Arsenate from Aqueous Solution by Nanocrystalline Mg/Al Layered Double Hydroxide: Sorption Characteristics, Prospects, and Challenges." *Water Sci. Technol.*, 61 (2010), 1411-1417.
8. A. Welch, S. Ryker, D. Helsel, and P. Hamilton, "Arsenic in Ground Water of the United States: An Overview." *Water Well J.*, 2001 (2001), 30-33.
9. C. Han, H. Li, H. Pu, H. Yu, L. Deng, S. Huang, and Y. Luo, "Synthesis and Characterization of Mesoporous Alumina and Their Performances for Removing Arsenic(V)." *Chem. Eng. J.*, 217 (2013), 1-9.
10. G. Zhang, A. Khorshed, and J. Paul Chen, "Simultaneous Removal of Arsenate and Arsenite by a Nanostructured Zirconium–Manganese Binary Hydrous Oxide: Behavior and Mechanism." *J. Colloid Interface Sci.*, 397 (2013), 137-143.
11. L. Ma and S. Tu, "Arsenic Removal from Water Using a Modified Rutile Ore and the Preliminary Mechanisms." *Desalin. Water Treat.*, 32 (2011), 445-452.
12. L. Yang, Z. Shahrivari, P.K.T. Liu, M. Sahimi, and T.T. Tsotsis, "Removal of Trace Levels of Arsenic and Selenium from Aqueous Solutions by Calcined and Uncalcined Layered Double Hydroxides (Ldh)." *Ind. Eng. Chem. Res.*, 44 (2005), 6804-6815.
13. K. Hristovski, P. Westerhoff, and J. Crittenden, "An Approach for Evaluating Nanomaterials for Use as Packed Bed Adsorber Media: A Case Study of Arsenate Removal by Titanate Nanofibers." *J. Hazard. Mater.*, 156 (2008), 604-611.
14. D. Vu, X. Li, Z. Li, and C. Wang, "Phase-Structure Effects of Electrospun Tio2 Nanofiber Membranes on As(III) Adsorption." *J. Chem. Eng. Data*, 58 (2013), 71-77.
15. E. Wahbah, *Removal of Arsenate and Arsenite by Calcined Nanocrystaline Layered Double Hydroxides*, 2015, (Master's Thesis, Department of Metallurgical Engineering, University of Utah).
16. N. Kondamudi, S.K. Mohapatra, and M. Misra, "Quintinite as a Bifunctional Heterogeneous Catalyst for Biodiesel Synthesis." *Applied Catalysis A: General*, 393 (2011), 36-43.
17. K.-H. Goh, T.-T. Lim, and Z. Dong, "Application of Layered Double Hydroxides for Removal of Oxyanions: A Review." *Water Research*, 42 (2008), 1343-1368.
18. S.J. Allen, G. McKay, and J.F. Porter, "Adsorption Isotherm Models for Basic Dye Adsorption by Peat in Single and Binary Component Systems." *J. Colloid Interface Sci.*, 280 (2004), 322-333.
19. F. Eba, S. Gueu, A. Eya'A-Mvongbote, J. Ondo, B. Yao, J.N. Nlo, and R.K. Biboutou, "Evaluation of the Absorption Capacity of the Natural Clay from Bikougou (Gabon) to Remove Mn (II) from Aqueous Solution." *Int. J. Eng. Sci. Tech.*, 2 (2010), 5001-5016.

Rare Metal Technology 2016
Edited by: Shafiq Alam, Hojong Kim, Neale R. Neelameggham, Takanari Ouchi, and Harald Oosterhof
TMS (The Minerals, Metals & Materials Society), 2016

EXPERIMENTAL STUDY ON VALUABLE METALS DISSOLUTION FROM COPPER SLAG

Sun Ying[1], Zhang Jing[1], Wang Yan-ze[1], Li Qiu-ju[1]

[1]Shanghai Key Laboratory of Modern Metallurgy & Materials processing, Shanghai University, shanghai 200072, China

Keywords: Copper slag, Valuable metals, Atmospheric sulfate leaching

Abstract

Copper smelter slag was leached using sulphuric acid solutions under atmospheric conditions. Base on the composition and phase of copper slag, the effect of the concentration of sulfuric acid and leaching temperature to the leaching growth rate of Fe, Cu, Co and Ni in the copper slag were investigated in this paper. Simultaneously, the influence of sulfuric acid concentration, leaching time, leaching temperature and the amount of sulfuric acid (solid-liquid ratio) to the leaching rate of these metal elements were researched. Experimental results showed that: with the increasing temperature, the dissolution rates of the metals elements increased first and then decreased. Due to the influence of silica gel which was obtained in beginning of leaching, the higher the concentration of sulfuric acid is not conducive to dissolution of the valuable metal elements. The dissolution degree of Fe, Cu, Co and Ni were 92.83%、96.76%、93.62% and 63.41% respectively.

Introduction

As one of the important raw materials in the development of the national economy, copper is widely used in the field of mechanical manufacturing, electronics, light industry, construction industry, national defense industry and other fields, which is the second largest non-ferrous metal. Copper slag with high-content copper is produced in the copper pyrometallurgical process. The copper content in the copper slag arrived to 0.5%. As a relatively poor country of copper resources, China's copper resources grade in the 0.2%~0.3% have been mined now[1]. More than 10 million tons of copper slag (containing nearly 100000 tons of copper) are produced from smelting copper in China every year[2]. The slag has the characteristics of large quantity, fine granularity, various types and complex composition and so on. Because of the considerable amounts of Ni, Co and other valuable metals contained in copper slag, it will result in waste of resources and serious pollution of the environment if copper slag is piled up for a long time.

Valuable metals can be extracted from copper slag integrally with Hydrometallurgical method. Compared with pyro-metallurgy, the energy consumption of hydro-metallurgy will be lower if the water does not have to be evaporated. What is more, the low content of the valuable metals can be extracted with hydrometallurgy. Leaching metal elements in the copper with hydrometallurgy is one of the most popular topics at home and abroad. A lot of research has been done in this area.

Muravyov, etc.[3] leached copper and zinc from copper converter slag flotation tailings using H_2SO_4 and biologically generated $Fe_2(SO_4)_3$. The recoveries of Cu and Zn were 81.6% and 37.7% respectively. Ahmed, etc.[4] leached and recovered copper and zinc by sulfuric acid leaching from copper slag, the two recoveries were 99% and 95%. The copper slag produced in the process of zinc smelting was treated with sulfuric acid by Zhang Zhongyi, etc.[5]. The effects of sulfuric acid concentration, leaching time, solid-liquid ratio and pressure on the leaching and enrichment of copper were studied in the experimental conditions. The results shown that the leaching rate of copper in copper slag was 98.1%.And crude copper enriched slag grade of more than 70% was obtained through purifying copper leaching solution.

Copper smelter slag was leached using sulphuric acid solutions under the condition of the atmosphere in this paper. The effects of sulfuric acid concentration, leaching time, leaching temperature and sulfuric acid dosage on the leaching rates of Fe, Co, Cu and Ni were studied in the experimental conditions. The experimental conditions were provided for improving the recovery rate of valuable metals and the resource utilization of copper slag.

Experimental

Materials and characterization

Fig.1 showed the chemical compositions and mineral phase of copper slag which was characterized through XRF methods. The chemical composition of the copper smelter slag was listed in Table 1. The compositions of copper in the form of element by calculating contained approximately 40 wt.% Fe, 5.11 wt. % Cu, 1.53 wt.% Zn.

Table 1 The composition and content of elements of the copper slag/ wt%

Elements	Fe	O	Si	Al	Cu	Zn	Pb	Co	Ni	others
Contents	39.98	33.06	17.24	0.94	5.11	1.53	1.39	0.15	0.02	0.58

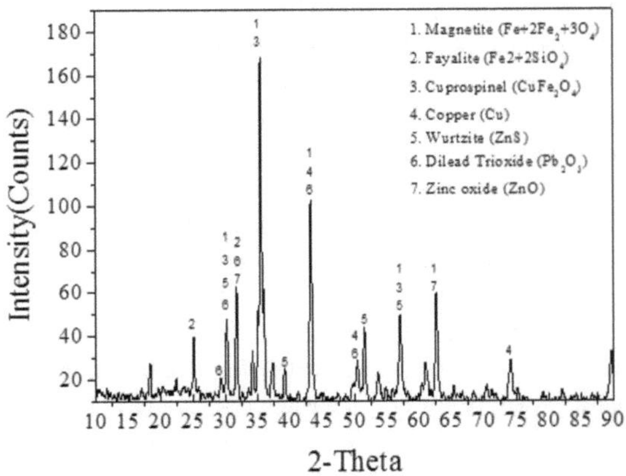

Fig. 1 XRD pattern of the copper slag

Fig.1 showed the phase composition of copper slag. The XRD pattern showed that the main phases of the copper slag were Fe_3O_4, Fe_2SiO_4 and copper-containing phase. In addition, a small amount of ZnO, Pb_2O_3 and other valuable metal compound were contained. Copper element mainly existed in the form of elemental metal. A small part of them were contained in the magnetite and sulfides. The content of Fe_2SiO_4 was about 60% by calculating.

Experimental processes

Experiments were carried out in 500ml beaker with a certain concentration of sulfuric acid. When reaching the test temperature by heating them in water bath, 10g copper slag was added to the beaker. During the experiment, leaching solution was repeatedly sampled. The concentrations of Fe, Cu, Co and Ni were obtained through ICP test. The particle size of the copper slag was 120 meshes (124μm) and the concentration of sulfuric acid is percentages quality.

Results and Discussion

Effect of leaching temperature

The effect of leaching temperature to metal leaching rates was investigated. This set of experiments was conducted concentrated sulfuric acid (30%) at a ratio of ore/ acid with 1:20 for 2 h. Leaching temperature were selected at 50℃, 60℃, 70℃, 80℃ and 90℃ respectively. The results were shown in Fig. 2.

As shown in Fig.2, the leaching rates of the metals increased as the leaching temperature was increased. But compared to other metals, the leaching rate of Ni was lower. At 90℃, the leaching rates of Fe, Cu, Co and Ni reached 94.5%, 69.7%, 73.9% and 36.4% respectively. So it was possible to say that the best leaching rates of the elements were obtained at the leaching temperature of 90℃.

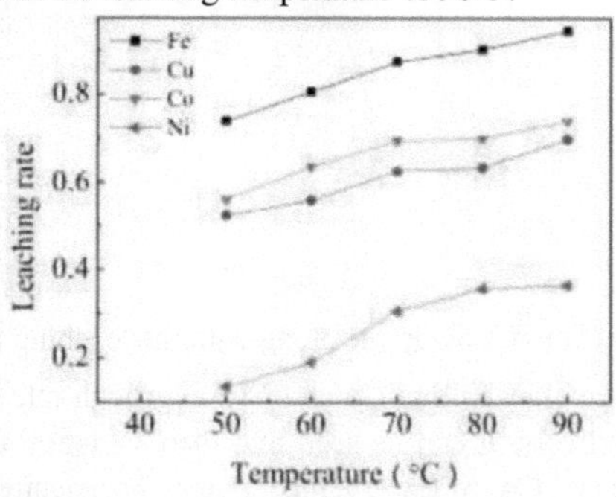

Fig. 2 Leaching rate of the metals along with leaching temperature

Considering that the leaching solution was taken several times, solid-liquid ratio was raised to 1:30 to reduce the impact of the liquid on the system. The effect of leaching temperature on leaching growth rate was investigated and shown in Fig.3.

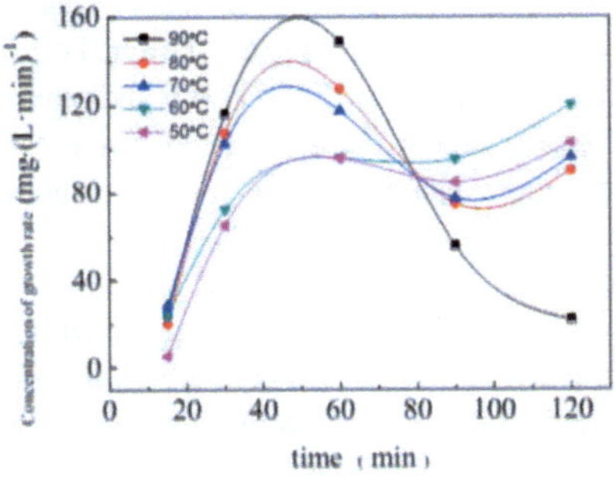

Fig. 3 Fe ion growth rate along with the leaching temperature

The leaching growth rate of Fe ion rapidly increased first and then slowed down. With the increase of the leaching temperature, it decreased faster and faster. It reached the highest value at 90℃.

Copper in the copper slag was dissolved with sulphuric acid. The dissolved reduction of Cu at different temperature was shown in Fig.4. Cu ion growth rate was increasing at 50℃. When the leaching temperature was 60℃, 70℃ or 80℃, the growth rate changed slightly in a certain period of time but was gradually increased on the whole. It reached the maximum at 60 min and then decreased at 90℃.

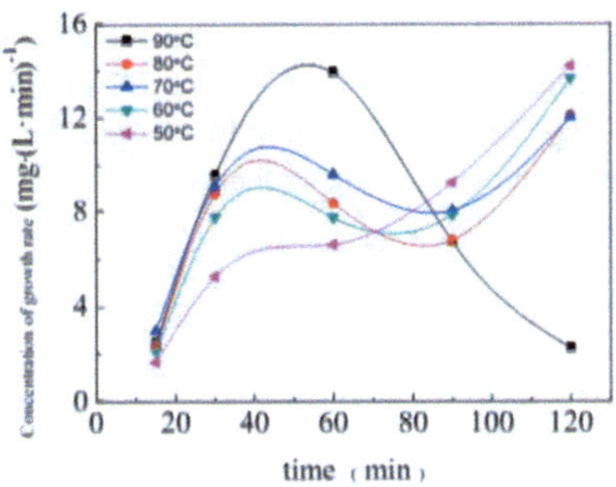

Fig. 4 Cu ion growth rate along with the leaching temperature

Fig.5 and Fig.6 showed that Co and Ni ion growth rate along with the leaching temperature respectively. It can be seen that their variation tendency were basically consistent with Fe and Co. At lower temperatures, the leaching growth rates increased first and then closed to balance. At higher temperatures, they increased significantly when the leaching time reached 30~60min and then slowed down gradually. What's more, leaching growth rates decreased faster with the increase of the leaching temperature.

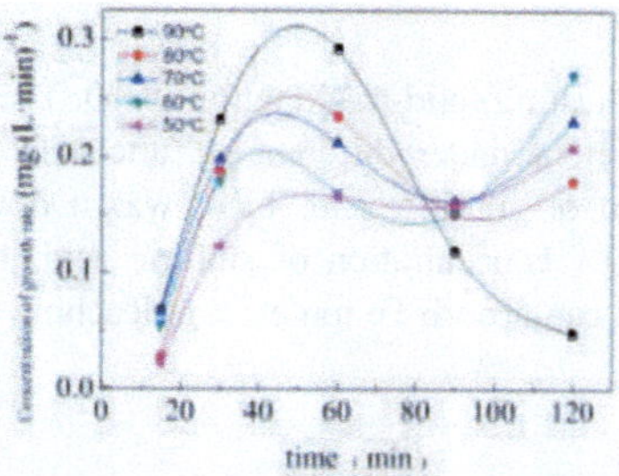

Fig. 5 Co ion growth rate along with the leaching temperature

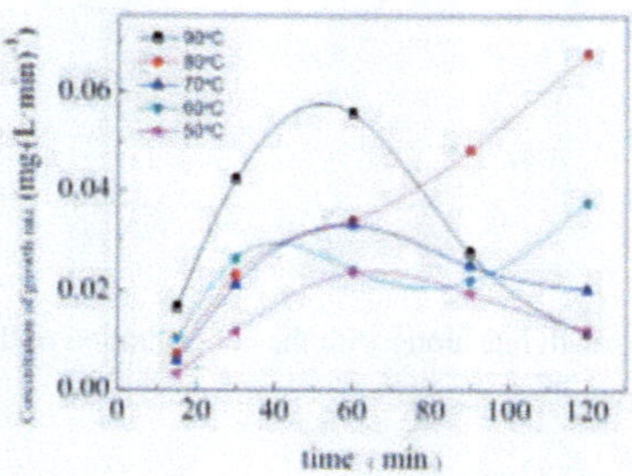

Fig. 6 Ni ion growth rate along with the leaching temperature

Effect of sulfuric acid concentration

The effect of sulfuric acid concentration metals leaching rates was investigated. This set of experiments was conducted an ore to acid ratio of 1:20 for 2h at 80℃.The sulfuric acid concentration was 10%, 20%, 30%, 40% and 50% respectively. The results of metals in copper slag dissolved were shown in Fig.7.

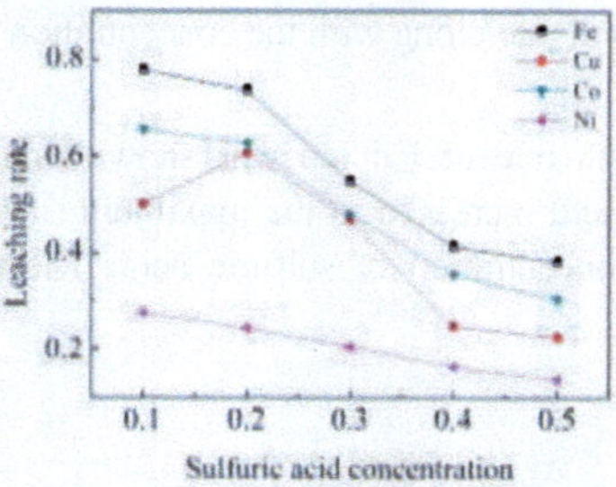

Fig. 7 Leaching rate of the metals along with sulfuric acid concentration

As shown in Fig.7, the leaching rates of Fe, Co and Ni reduced with the increasing concentration of sulfuric acid. The leaching rate of Cu increased slightly in low concentration of sulfuric acid and decreased with the sulfuric acid concentration over 20%. Considering silica gel would be prepared later, sulfuric acid concentration of 20% was appropriate.

Fig.8 showed the effect of sulfuric acid concentration on leaching growth rate was investigated with adjusting solid-liquid ratio to 1:30. The leaching growth rates of Fe ion were quite different under different sulfuric acid concentration. With the increasing concentration of sulfuric acid, there was a downward trend and were negative numbers at 50% concentration of sulfuric acid. It can be inferred that a certain substance which can absorb Fe ion exist in leaching system. It was proved to be silica gel.

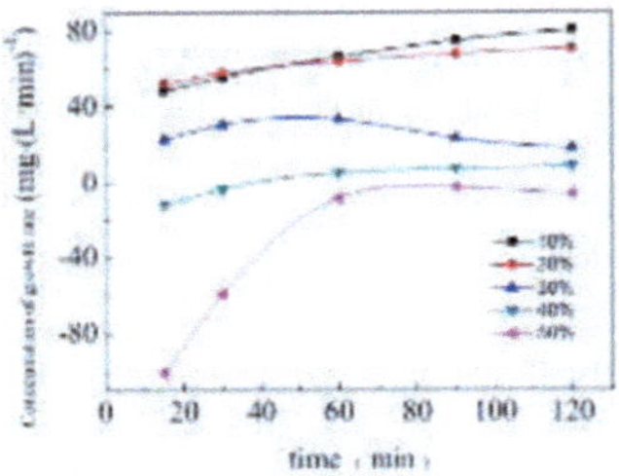

Fig. 8 Fe ion growth rate along with the concentration of the sulfuric acid

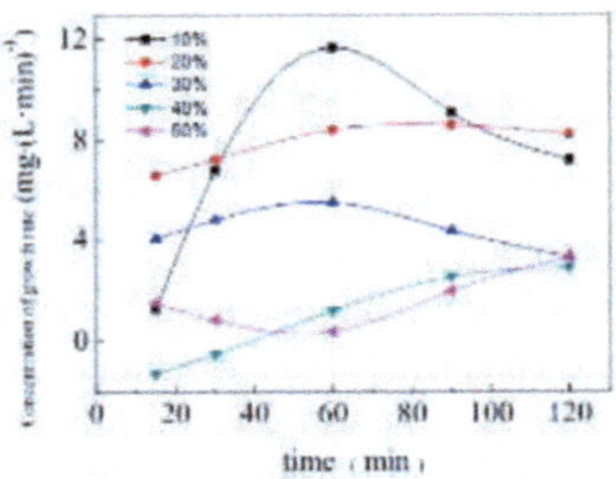

Fig. 9 Cu ion growth rate along with the concentration of the sulfuric acid

The leaching growth rate of Cu ion was shown in Fig.9. Cu in copper slag dissolved with sulfuric acid increased to the maximum first and then decreased to a certain value at 10% concentration of sulfuric acid. The maximum value reached $12mg\cdot L^{-1}\cdot min^{-1}$.

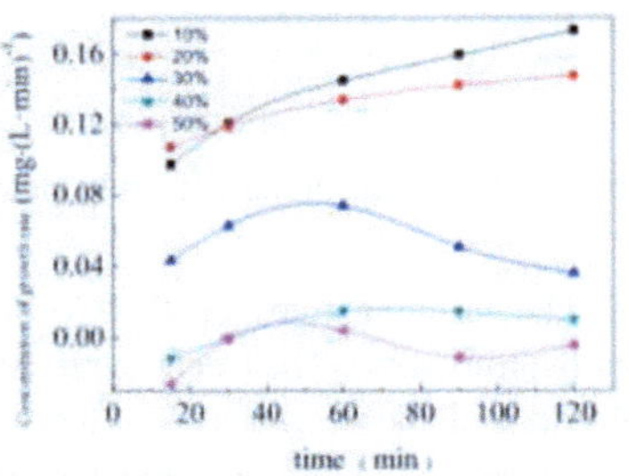

Fig. 10 Co ion growth rate along with the concentration of the sulfuric acid

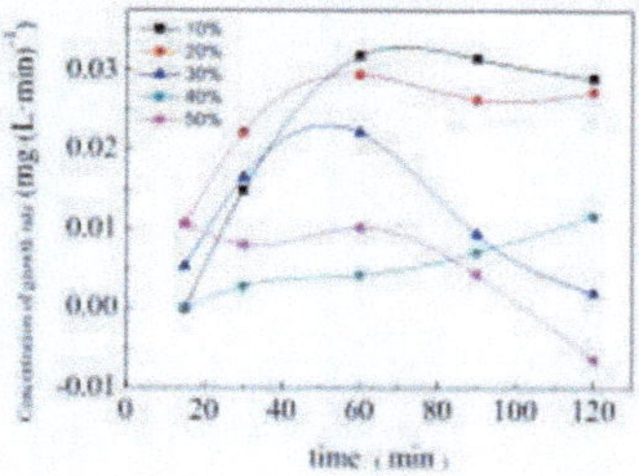

Fig. 11 Ni ion growth rate along with the concentration of the sulfuric acid

Fig.10 and Fig.11 showed that Co and Ni ion growth rate along with the concentration of the sulfuric acid respectively. The leaching growth rates of the two elements showed a downward trend and then even a negative growth with the increasing concentration of sulfuric acid. This showed that silica gel can absorb each metal ion and the high concentration of sulfuric acid may promote the formation of silica gel.

Effect of leaching time

The effect of leaching time to metal leaching rates was investigated. This set of experiments was conducted concentrated sulfuric acid (30%) at an ore to acid ratio of 1:20 at 80℃.The leaching time was 15min, 30min, 60min, 90min and 120min respectively. The results are shown in Fig.12.

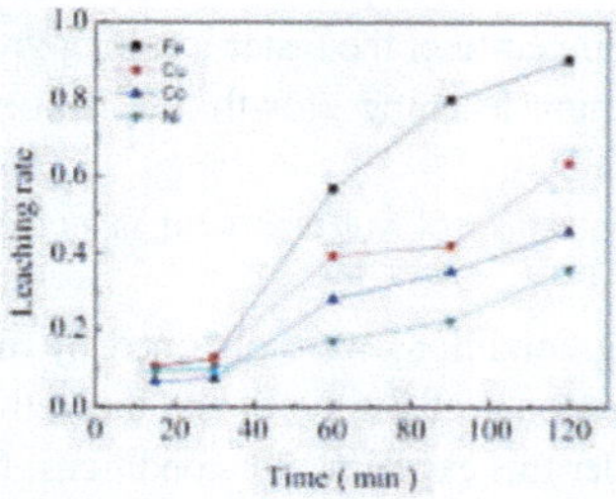

Fig. 12 Leaching rate of the metals along with leaching time

Fig.12 indicated that the leaching rates of the metal ions increased with the increasing leaching time. They were litter difference from 15min to 30min. After 30min, the leaching rate of Fe ion increased significantly, while the leaching rate of Cu, Co and Ni increased slowly. When the leaching time reached 2h, the leaching rates of the metal ions changed gradually. So the leaching time was 2h.

Effect of the amount of sulfuric acid

The effect of the amount of sulfuric acid to metal leaching rates was investigated. This set of experiments was conducted concentrated sulfuric acid (30%) at 80℃ for 2h.The amount of sulfuric acid was 1, 3, 5, 7 and 9 times of theoretical amount respectively. The results are shown in Fig.13.

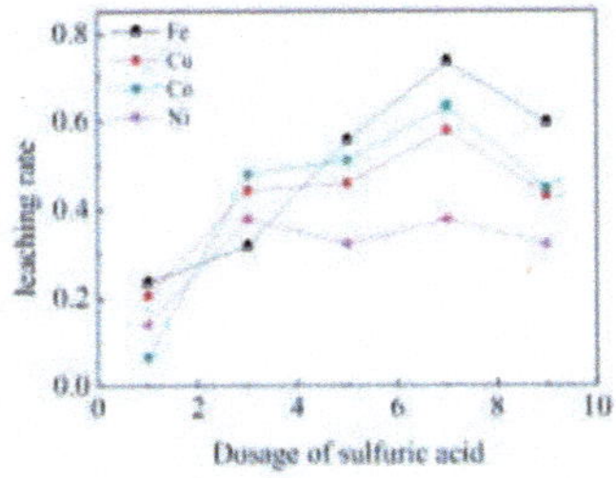

Fig. 13 Leaching rate of the metals along with Dosage of sulfuric acid

It can be seen that the leaching rates of Fe, Cu, Co and Ni increased significantly initially and then showed a downward trend with the increase of the amount of sulfuric acid. This may be the result of the experiment without stirring. The leaching rates of the metal elements reached maximum then the dosage of sulfuric acid was 7 time of the theoretical amount.

In view of the foregoing results, the optimum conditions is to mix the copper slag with 30% sulfuric acid in amount of 7 time of the theoretical amount for 2h leaching time at 90℃. Under this experimental conditions, the dissolution degree of Fe, Cu, Co and Ni were 92.83%、96.76%、93.62% and 63.41% respectively.

Conclusion

This work is devoted to the study on valuable metals dissolution from copper slag by using sulphuric acid solutions under the condition of the atmosphere. The results obtained from the laboratory experiments can be summarized as follows:

(1) The higher leaching temperature, the faster leaching growth rate of the metal ions, and the greater maximum leaching growth rate. But after that, the more rapidly leaching growth rate fall.

(2) The higher the concentration of sulfuric acid is not conducive to the metal ion leaching.

(3) The optimum leaching conditions were obtained by mixing the copper slag with 30% sulfuric acid in amount of 7 time of the theoretical amount for 2h leaching time at 90℃. And under this experimental conditions, the dissolution degree of Fe, Cu, Co and Ni were 92.83%、96.76%、93.62% and 63.41% respectively.

References

[1] Ministry of Land and Resources of the People's Republic of China. China Mineral Resources Report[R]. Beijing: China Geological Publishing House, 2013: 7-20.

[2] ZHU Zu-ze, HE Jia-qi. The modern metallurgy of copper[M]. Beijing: Science Press, 2003: 12-18.

[3] Muravyov M I, Fomchenko N V, Usoltsev A V, et al. Leaching of copper and zinc from copper converter slag flotation tailings using H_2SO_4 and biologically generated $Fe_2(SO_4)_3$[J]. Hydrometallurgy, 2012, 199:40-46.

[4] Ahmed I M, Nayl A A, Daoud J A. Leaching and recovery of zinc and copper from brass slag by sulfuric acid[J]. Journal of Saudi Chemical Society, 2012.

[5] Zhang Zhong-yi, Kuang Zhi-en, Yang Gang. Experimental study on the recovery of Zn and Cu in copper slag[J]. Yunnan Metallurgy, 2008, 37(1):27-29.

Rare Metal Technology 2016
Edited by: Shafiq Alam, Hojong Kim, Neale R. Neelameggham, Takanari Ouchi, and Harald Oosterhof
TMS (The Minerals, Metals & Materials Society), 2016

Adsorption of Platinum and Palladium from Hydrochloric Acid Media by Hydrothermally Treated Garlic Waste Gel

Bo Liang[1], Kai Huang[1,*], Hongmin Zhu[1], Shafiq Alam[2]

[1]School of Metallurgical & Ecological Engineering, University of Science and Technology Beijing, Beijing, Xueyuan Rd.30 Haidian District, 100083, China
[2]Department of Chemical and Biological Engineering, University of Saskatchewan, Room 1C123 Engineering Building, Canada

*Corresponding author: khuang@metall.ustb.edu.cn; +86-10-62334204

Keywords: Platinum, Palladium, adsorption, garlic peel, hydrothermal ammoniation.

Abstract

A new approach of recovering platinum and palladium from electronic wastes by hydrochloric acid leaching followed by selective adsorption on low-cost and environmentally benign biomass sorbent prepared from easily available agriculture waste is suggested. The obtained adsorbents were prepared by a combination of chemical modification and hydrothermal carbonization process with $(NH_4)_2C_2O_4$. It was found that the maximum adsorption capacity of hydrothermal carbonized garlic peel for Pt(IV) and Pd(II) was 1.64 and 0.42 mol/kg respectively. The Langmuir adsorption isotherm was applied to describe the adsorption process. Kinetics of the Pt(IV) and Pd(II) ions adsorption was found to follow pseudo-second-order rate equation. The adsorbents were examined by scanning electron microscope (SEM), and Fourier transform infrared spectroscopy (FT-IR). Based on the characterization results, a possible mechanism of Pt(IV) and Pd(II) ions adsorption with the hydrothermal carbonized garlic peel was proposed. This study provides a promising adsorbent with efficient adsorption property for platinum and palladium recovery from the leach solution of electronic wastes.

Introduction

Platinum and palladium are important precious metals for its specific physical and chemical properties, and they have been historically essential as currency. The global demand for them has increased rapidly in recent years. But the deposit amount of natural resources is depleting due to the continuous mining and consumption. Consequently, the cost of platinum and palladium production and its price have risen sharply[1]. On the other hand, cheaper and more advanced electronic gadgets are continuously replacing the old ones as a result of rapid technological advancement in modern society[2]. Electrical and electronic wastes contain high value precious metals like gold, platinum and palladium[3]. Although, the portion of these metals is very low compared to other base metals like copper and iron in such wastes, their contents in this form are much higher than those in their original ores[4].

To date, many techniques have been developed to recover the valuable metals from various e-wastes, and hydrometallurgical process has drawn much attention due to its simpleness, effectiveness, and eco-friendliness during the extractive process, in which the most challenge is how to recover these low concentration (usually around several and several ten ppms) valuable metal ions from the leaching solution effectively. The common methods include ion exchange by resins, precipitation, and solvent extraction[5-12]. However, these techniques usually suffer from some drawbacks such as being expensive. Over the last two decades, adsorption has been considered to be a promising technology for the removal and/or recovery of metal ions with the advantages of low cost and simple operations[13-18]. Obviously, the preparation of cost-effective adsorbents is the key to the realization of the above ideas, and especially the biosorbents synthesized from agricultural waste biomasses have got great attraction. The common chemical modification methods to synthesize the functional materials for adsorption are complex, tedious and producing excessive residual reagents, so the facile synthesis methods are quite fascinating, and hydrothermal carbonization is exactly advantageous in these aspects for biomass treatment and converting.

Hydrothermal carbonization (HTA) technology is a "green" and versatile strategy for the synthesis of advanced carbon materials suitable for a wide variety of applications in many industries[14]. In an HTA process, pure carbohydrates as well as raw lignocellulosic biomass wastes can be converted into valuable carbon materials using water as a carbonization medium at mild temperature (< 200 ℃) under self-generated pressure[15]. As studied by Titirici, the formation of HTA-based materials is a very complex process due to the multiple simultaneous reactions occurring in the hydrothermal treatment, probably including dissolution, dehydration, polymerization, and aromatization reactions, and the obtained HTA products usually are composed of the condensed furan rings bridged by aliphatic regions with terminal hydroxyl and carbonyl functional groups[19,20]. By changing the processing temperature, the compositions and morphology of the product may have quite differences, and so it will have different adsorption behavior for metal ions. Recently, several biomasses have been used for adsorptive recovery of gold including tannin, fungal biomass, alfalfa, various protein sources and fruit wastes[16-21], though the biosorbents prepared by these biomasses show good affinity to gold ions, their sources are a little specially required. If the ordinarily common biowastes can be used to prepare the similar biosorbents for the gold capturing, it would be greatly meaningful to the development of cheaper and smarter biosorbents. So for this purpose, we would like to apply an easy one-step synthesis method, i.e., the hydrothermal carbonization technology, to treat the garlic peel waste in present study. Garlic peel, as a kind of easily available agricultural waste because of large consumption of garlic in daily life can be chosen as the raw material to synthesize the alternative adsorbent for gold recovery process. As well known, garlic peel is a typical lignocellulosic biomass rich in hydroxyl, carboxyl, sulfate, and amino groups, which may bond metals ions.

To the best of our knowledge, there is no report on the recovery of platinum and palladium using garlic peel modified by hydrothermal carbonization. The main objective of the present work is to investigate the properties of the hydrothermal carbonized garlic peel as a new adsorbent for the recovery of platinum and palladium, and during our study it was found to exhibit excellent adsorption capacity for platinum and palladium with high efficiency and quick adsorption. Therefore, it is promising to apply this new cheap and effective adsorbent for the recovery of platinum and palladium from electrical and electronic wastes leaching liquor with hydrochloric acid as the lixiviant.

2. Materials and Methods

2.1 Reagents and Preparation of HTA-GP

The reagents used for the preparation of hydrothermal carbonized garlic peel include calcium hydroxide (CaO, 98%) purchased from Aladdin Company and distilled water. The garlic peel was supplied by a factory in Shandong province in China. Chloroplatinic acid hexahydrate ($H_2PtCl_6\ 6H_2O$, 99%) and palladium chloride ($PdCl_2$, 99%) from Beijing Chemical Works were used in the single and multi-components adsorption experiments. The multi-components solutions also contain iron (III) nitrate nonahydrate ($Fe(NO_3)_3\ 9H_2O$,98%), copper nitrate trihydrate ($Cu(NO_3)_2\ 3H_2O$),99%), nickel nitrate hexahydrate ($Ni(NO_3)_2\ 6H_2O$,98%) and zinc nitrate hexahydrate ($Zn(NO_3)_2\ 6H_2O$, 99%) from Aladdin Company. Standard calibration solutions of 1000 mg/L gold in 2% hydrochloric acid and 1000 mg/L copper, iron, nickel, zinc in 2% nitric acid were purchased from China General Institute of Steel and Iron Company.

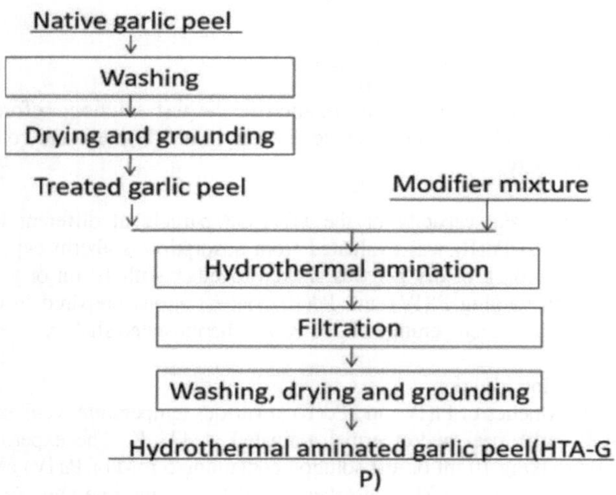

Scheme I: Flowsheet of HTA-GP gel by hydrothermal ammoniation of garlic peel

The native garlic peel was washed several times with tap water and dried in an oven at 60 ℃ for 24 h. The dried material was ground and sieved to below the desired mesh size of 150 micrometers. Fifteen grams dried native garlic peel and 5 gram of CaO powder was crushed together and dipped in 250 ml water, stirred for 24 h at 25 ℃. Then, it was filtered and washed with distilled water up to pH 7 and dried in an oven at 100 ℃ for 12 h and then kept in a desiccators, which is called saponified garlic peel (S-GP).

The prepared garlic peel was hydrothermally aminated by mixing 1 g of S-GP garlic peel with 15 ml ammonium compounds solution in a 30 ml autoclave with a Teflon lining for 24 h at 110℃, 150℃ and 200℃, respectively. Then it was taken out, filtered and washed with distilled water up to pH 7 and ground, sieved to below 100 mesh after dried in an oven at 100 ℃ for 12 h, and the obtained powder is called HTA-GP. The adsorbents before and after adsorption were characterized Fourier transform infrared spectroscopy (FTIR, Perkin-Elmer GX 2000) to determine chemical properties.

2.2 The adsorption isotherms

All the adsorption studies were carried out in batch modes. Briefly, sample solutions containing designed metal concentration were mixed with hydrothermal aminated garlic peel and shaken in a thermostated shaking incubator. After adsorption, the samples were filtered immediately and the metal ion concentration in the filtrate was quantitatively measured by inductively coupled plasma, atomic emission spectrometer (ICP-AES, Perkin Elmer Optima 3000XL) and/or atomic absorption spectrophotometer (AAS, Shimadzu AA 6800). The percentage adsorption (A) and the amount of metal adsorbed (q) on the adsorbent was calculated as follows:

$$A = \frac{C_i - C}{C_i} \times 100 \quad (1)$$

$$Q = \frac{C_i - C}{W} \times V \quad (2)$$

Where C_i and C are the metal concentration in the test solution, before and after adsorption, respectively, W and V denote dry weight of adsorbent and volume of test solution, respectively.

The maximum uptake capacity of the HTA-GP particle at different temperature towards Pt(IV) and Pd(II) was evaluated from adsorption isotherm experiments. In this experiment, 0.02 g of dry gel was shaken together with 10 ml of test solutions individually with varying Pt(IV) and Pd(II) concentrations prepared in 0.1 M HCl solution for 48 h at the temperature of 298 K in a thermostated shaker.

2.3 The adsorption kinetics

The adsorption kinetics of Pt(IV) and Pd(II) at various temperature were examined for the HTA-GP garlic peel hydrothermal aminated at 473 K. The experiments were conducted by shaking 10 ml of test solution containing 5 mM of Pt(IV) and Pd(II) in 0.1 M HCl together with 0.02 g adsorbent for definite time intervals. At the end of each contact time, the aliquots of the suspended solutions were filtered and the metal concentration of the filtrate was measured as described above.

3. Results and discussion

3.1 Surface morphology of HTA-GP gel

Fig.1 shows the microstructure of the garlic peels before and after HTA treatment, and it can be found that after modification at high temperature, quite a few components of the garlic peel were lost in the form of water soluble species, which is probably dissolved in the aqueous solution. The residual of solid, i.e., HTA-GP gel, looks like fiber ring.

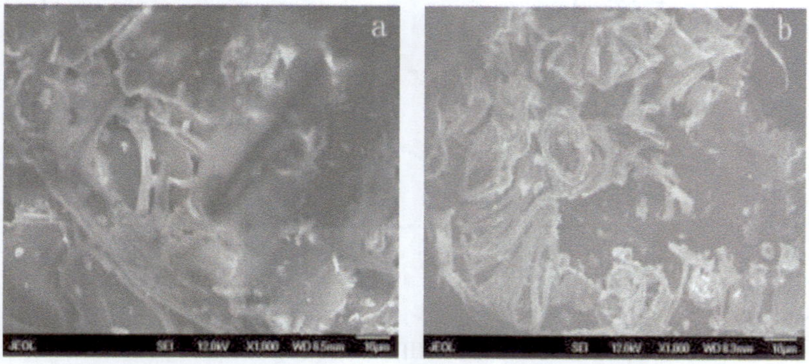

Fig.1 SEM graphs of garlic peel before and after HTA treatment

3.2 Screening of suitable chemical reagent for modification

The aim of modification is to graft $-NH_2$ ligands onto the garlic peel, and makes it have strong affinity to the soft metals of Pt and Pd. In this study, four kinds of soluble compounds of ammonia, i.e., ammonium oxalate, ammonium chloride, ethylenediamine and phenylamine, were used to treat the garlic peel and the adsorption capacity of the obtained gels were evaluated and listed in **Table 1**. It can be seen that ammonium oxalate is the best choice due to its both excellent adsorption affinity for Pt(IV) and Pd(II). The suitable temperature of hydrothermal treatment was evaluated at three points, and the obtained gels were tested for their adsorption capacity of gold ions, and the results were demonstrated in Fig.2, and it can be found that the higher treatment temperature id favorable for the adsorption of gel. So the gel obtained at 200℃ is recommended in this study.

Table 1 Maximum adsorption capacity of HTA-GP gel obtained by using different ammonium salts as the modifier

Modifiers	Q(mol/kg)	
	Pt	Pd
Ammonium oxalate	1.64	0.42
Ammonium chloride	1.04	0.14
Ethylenediamine	0.70	0.35
Phenylamine	0.87	0.62

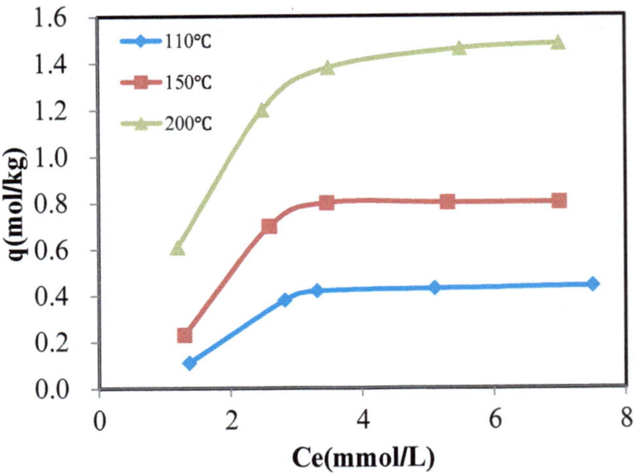

Fig.2 Adsorption isotherm of Au(III) onto HTC-garlic peel gel obtained under different hydrothermal temperatures

3.3 Adsorption isotherms of Pt(IV) and Pd(II) on HTA-GP

The maximum metal uptake capacity of the garlic peel hydrothermally aminated at varying temperatures was evaluated in terms of the adsorption isotherms. The tests were examined at different initial Pt(IV) and Pd(II) concentrations of the test solution. **Fig.3** shows the results of adsorption isotherms of Pt(IV) and Pd(II) on different HTA-GPs. As seen in Fig.3, it is clear that the amount of Pt(IV) and Pd(II) adsorption on the adsorbent increased with increasing metal concentration of the test solution and tended to approach a constant value at the high concentration of metal ions. Further study shows that the above experimental data fit well with the Langmuir model, which indicates that the adsorption process can be described by the monolayer covering phenomena.

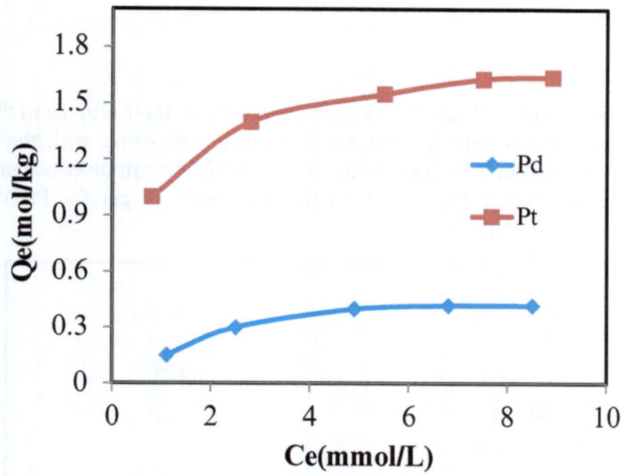

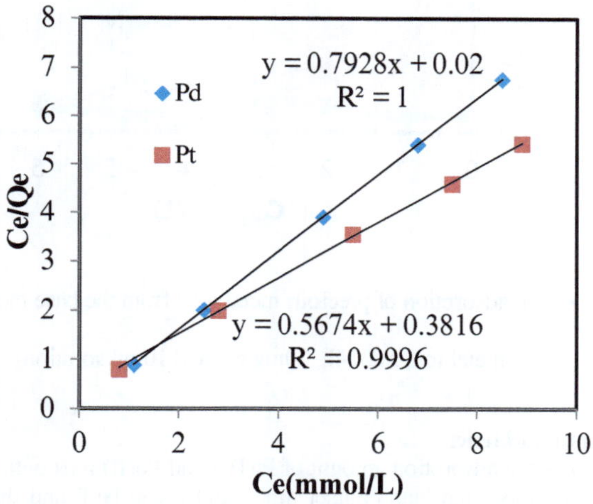

Fig.3 Adsorption isotherm of Pd(II) and Pt(IV) onto HTA-garlic peel gel (0.05 M HCl, r.t., 24h, 20mg gel and 10 ml solution) (upper figure) and its fitting curves based on Langmuir model(bottom figure)

3.4 Selective recovery of Pt(IV) and Pd(II) by HTA-GP gel

Fig.4 is the adsorption behavior of HTA-GP gel from the mixed solution containing the precious metal ions and base metal ions in the hydrochloric acid medium. It can be seen that the gel shows its selective adsorption property only at the very dilute HCl solution. This result shows the inhibition effect of HCl, and tells that the leaching solution of the relative e-wastes generally needs dilution operation if the selective recovery of Pt and Pd is anticipated. At the same time, it should be noticed that the chemical compounds of base metals were adopted with their nitrates in this study, and their concentration is very low (about 2~3 mM), comparing with the hydrochloric concentration tested in the experiments (0.1M ~ 6M), the nitrate concentration is quite much lower, so its negative effect on the adsorption of gel for Pt(IV) and Pd(II) probably can be ignored.

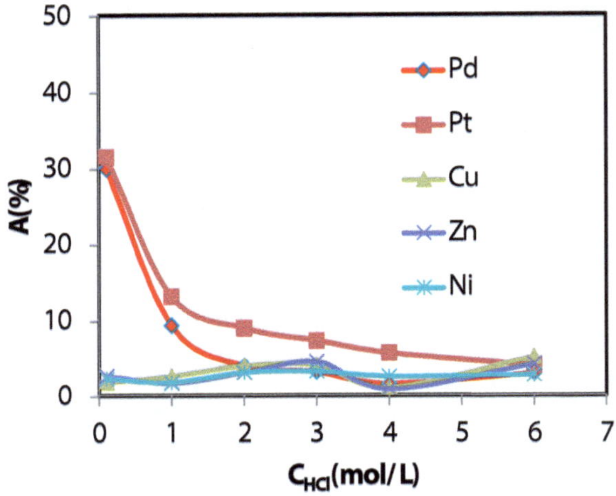

Fig.4 Selective adsorption of precious metal ions from the base metal ions(1 mM metal ion, r.t., 24h, 20mg gel and 10 ml solution)

3.5 Adsorption kinetics

Fig.5 indicates the adsorption amount of Pt(IV) and Pd(II) ions onto the HTA-GP gel at different temperature and contact time. And it can be found that the adsorption rates for both precious metal ions are quite rapid, and both attain equilibrium state after 5 h contact. It also can be seen that increasing of temperature is not beneficial to the adsorption of Pt(IV) and Pd(II) ions onto the HTA-GP gel , which suggests that the adsorption is an exothermic process, and so room temperature is recommended. Further study shows that the adsorption process both fit well with the pseudo-second order equations.

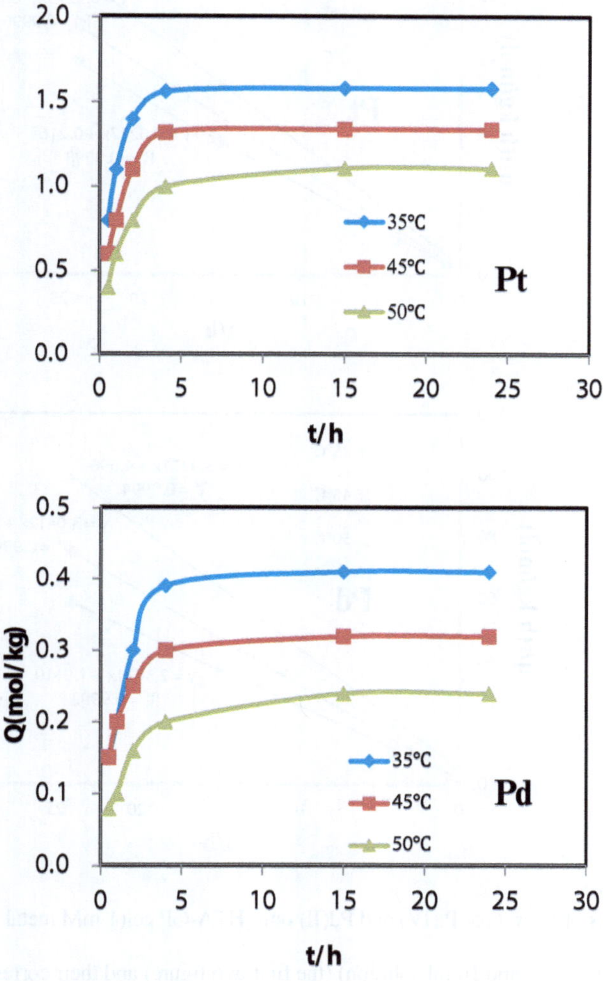

Pt
35°C
45°C
50°C
t/h
Pd
Q(mol/kg)

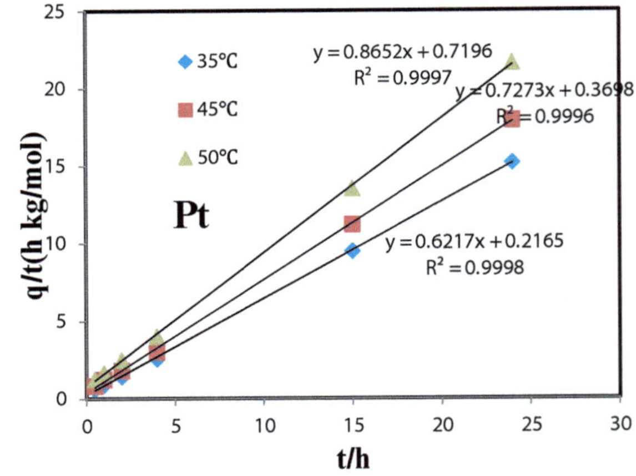

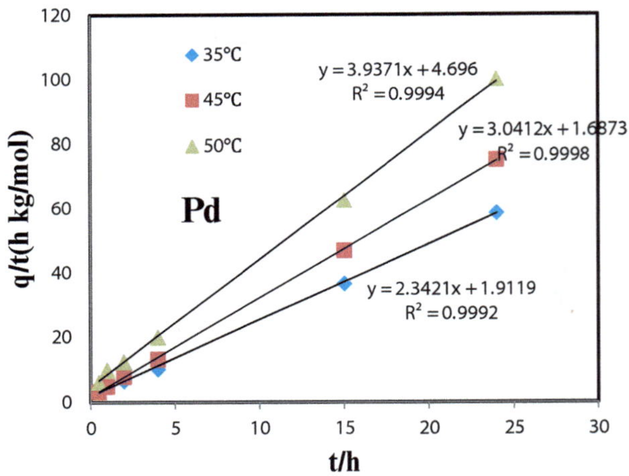

Fig.5 Plots of Q *vs.* t for Pt(IV) and Pd(II) onto HTA-GP gel(1 mM metal ion, 0.05 M HCl, 20mg gel and 10 ml solution) (the first two figure) and their corresponding pseudo-second-order kinetics fitting curves(the last two figures)

3.5 Adsorption mechanism

The native garlic peel was used to test its adsorption for Pt(IV) and Pd(II) ions, and there is no any adsorption efficiency, but after chemical modification by HTA treatment it shows adsorption property, which proved the $-NH_2$ ligands can be effectively grafted onto the garlic peel, which is regarded as the key role for the adsorption of Pt(IV) and Pd(II) ions. **Fig.6** is the FTIR results, and it can be found that after modification, the peaks at 1612 cm^{-1} and 1320 cm^{-1} become weaker which shows the $-NH_2$ ligands effectively fixed to the garlic peel surface. And after adsorption of Pt(IV) metal ions, the peak at 1120 cm^{-1} becomes weaken drastically showing the effective uptaking of Pt(IV). And at the high concentration of hydrochloric acid, the grafted $-NH_2$ ligands may be destroyed by strong acid, so leading to the decrease of the adsorption of precious metal ions. So the chloro-species of Pt(IV) and Pd(II), i.e., $PtCl_6^{2-}$ and $PdCl_4^{2-}$, are believed to be captured by the protonated $-NH_2$ through the electrostatic attraction force.

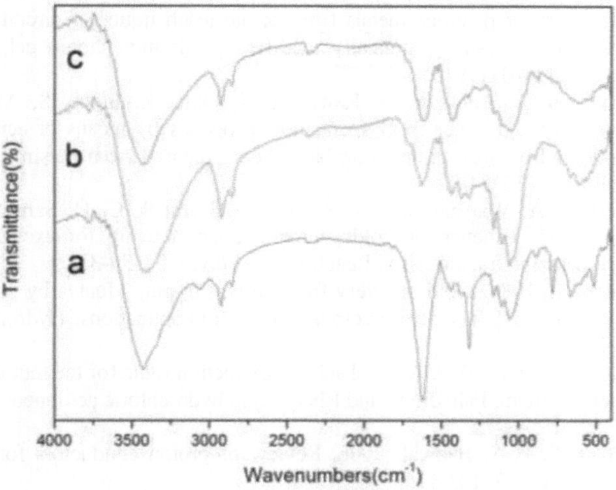

Fig.6 FT-IR spectra of garlic peel(a) native garlic peel,(b) HTA-GP, (c) after adsorption of Pt(IV).

Conclusion

The native garlic peel was successfully modified by the hydrothermal treatment in the ammonia oxalate solution. Adsorption of Pt (IV) and Pd(II) on HTA-GP gel followed a Langmuir isotherm and the maximum adsorption capacity for Pt (IV) and Pd(II) were 1.62 and 0.42 mol/kg. The adsorption was found to follow a pseudo-second ordered kinetic process and it needed about 4 hours to reach equilibrium

Acknowledgements

The authors want to show their thanks to the financial support from the Education Committee of Beijing City (Grant No.20140101), the Scientific Committee of Beijing City (Grant No.20130627).

References

1 K.Fujiware, A. Ramesh, T. Maki, H. Hasegawa, K. Ueda, 2007. Adsorption of platinum(IV), palladium(II) and gold(III) from aqueous solutions onto l-lysine modified crosslinked chitosan resin. J. Hazard. Mater. 146: 39-50

2 M. Gurung, B. B. Adhikari, H. Kawakita, K. Ohto, K. Inoue, S. Alam, 2011. Recovery of Au(III) by using low cost adsorbent prepared from persimmon tannin extract. Chem. Eng. J, 174, 556-563.

3 M. Gurung, B. B. Adhikari, H. Kawakita, K. Ohto, K. Inoue, S. Alam, 2012. Selective recovery of precious metals from acidic leach liquor of circuit boards of spent mobile phones using chemically modified persimmon tannin gel. Ind. Eng. Chem. Res., 51, 11901-11913.

4 M. Gurung, B. B. Adhikari, H. Kawakita, K. Ohto, K. Inoue, S. Alam, 2013. Recovery of gold and silver from spent mobile phones by means of acidothiourea leaching followed by adsorption using biosorbent prepared from persimmon tannin. Hydrometallurgy, 133, 84-93.

5 J. L. Cortina, A. Warshawsky, N. Kahana, V. Kampel, C. H. Sampaio, R. M. Kautzmann, 2003. Kinetics of goldcyanide extraction using ion-exchange resins containing piperazine functionality. React. Funct. Polym, 54, 25-35.

6 P. F. Sorensen, 1988. Gold recovery from carbon-in-pulp eluates by precipitation with a mineral acid III. The acid precipitation step in applications. Hydrometallurgy, 21(2), 249-254.

7 P. Malik, A. P. Paiva, 2010. A novel solvent extraction route for the mutual separation of Platinum, Palladium, and Rhodium in hydrochloric acid media. Solv. Extr. Ion Exch, 28, 49-72.

8 J. W. Phair, S. P. S. Badwal, 2006. Review of proton conductors for hydrogen separation. Ionics, 12, 103-115.

9 Y. H. Ma, I. P. Mardilovich, E. E. Engwall, 2003. Thin composite palladium and palladium/alloy membranes for hydrogen separation. Annals. New York Academy of Sciences, 984, 346-360.

10 K. L. Yeung, J. M. Sebastian, A. Varma, 1995. Novel preparation of Pd/Vycor composite membranes. Catal. Today, 25, 231-236.

11 K. L. Yeung, R. Aravind, J. Szegner, A. Varma, 1996. Metal composite membranes: synthesis, characterization and reaction studies. Stud. In Surf. Sci. Catal. 101, 1349-1358.

12 Y.S. Cheng, K. L. Yeung, 2001. Effects of electroless plating chemistry on the synthesis of palladium membranes. J. Membr. Sci., 182, 195-203.

13 W. S. W. Ngah, M. A. K. Hanafiah, 2008. Adsorption of copper on rubber (Hevea brasiliensis) leaf power: kinetic, equilibrium and thermodynamic studies. Biochem. Eng. J, **39**, 521-530.

14 A. Nakajima, T. Sakaguchi, 1993. Uptake and recovery of gold by immobilized persimmon tannin. J. Chem. Technol. Biotechnol., 57, 321-326.

15 A. Nakajima, Y. Baba, 2004. Mechanism of hexavalent chromium adsorption by persimmon tannin gel. Water Res., 38, 2859-2864.

16 T. Ogata, Y. Nakano, 2005. Mechanisms of gold recovery from aqueous solutions using a novel tannin gel adsorbent synthesized from natural condensed tannin. Water Res., 39, 4281-4286.

17 M. Yurtsever, I. A. Sengil, 2009. Biosorption of Pb(II) ions by modified quebracho tannin resin. J. Hazard. Mater., 163, 58-64.

18 X. Li, C. Zhang, R. Zhao, X. Lu, X. Xu, X. Jia, C. Wang, L. Li, 2013. Efficient

adsorption of gold ions from aqueous systems with thioamide-group chelating nanofiber membranes. Chem. Eng. J., 229, 420-428.

19 M. Gurung, B. B. Adhikari, S. Alam, H. Kawakita, K. Ohto, K. Inoue, 2013. Persimmon tannin-based new sorption material for resource recycling and recovery of precious metals. Chem. Eng. J., 228, 405-414.

20 M.M. Titirici, M. Antonietti, 2010. Chemistry and materials options of sustainable carbon materials made by hydrothermal carbonization. Chem. Soc. Rev, 39, 103-116.

21 O.A. Mohammad, A.W.T. Ivy, S.L. Leo, 2011. Automobile adsorption air-conditioning system using oil palm biomass-based active carbon: A review. Renew. Sust. Energ. Rev. 15, 2061-2072.

Rare Metal Technology 2016
Edited by: Shafiq Alam, Hojong Kim, Neale R. Neelameggham, Takanari Ouchi, and Harald Oosterhof
TMS (The Minerals, Metals & Materials Society), 2016

PRESSURE OXIDATION LEACHING OF GOLD-ANTIMONY ALLOY

Aichun Dou

School of material science and engineering, Jiangsu University, Zhengjiang 212013, China

Keywords: pressure, oxidation leaching, separation, Au-Sb alloy

Abstract

Au-Sb alloy sample in which the total amount of Cu and Ni is below 25% (wt. %). In practice, Cu and Ni were difficult to be dissolved in sulphuric acid solution by oxidative leaching under the atmospheric pressure conditions, because of the lower concentration of Cu and Ni in the sample. In this paper, pressure oxidation leaching in sulphuric acid system (POLS) followed by hydrochloric acid leaching (HAL) was employed to remove impurities in the Au-Sb alloy. The effects of variables in POLS on leaching ratios of different metals were investigated in detail. Under the optimal conditions: the concentration of H_2SO_4 is 1.5 mol/L, liquid-solid ratio is 7:1, the amount of oxidizer is 1.2 times of the theoretical amount, leaching time is 4h and leaching temperature is 427K, the leaching ratios of copper and nickel reached up to 99%, antimony, lead and silver were not dissolved in dilute sulphuric acid solution. The residue from POLS was then treated with hydrochloric acid solution in fixed conditions, the leaching ratio of antimony reached over 92%, the grade of gold in the ultima residue was enriched up to 66%.

Introduction

The gold-antimony concentrate is an important resource for extracting antimony and gold. Generally, the concentrate is firstly smelted in electric furnace [1] or in blast furnace [2] to enrich the precious metals, because the metallic antimony is a good collecting agent for gold. In practice, above 98% of gold in concentrate is collected in the melt of metallic antimony, the component of gold in the melt is low. While volatilized most of the antimony in the melt as crude antimony white, the component of gold would be higher. The remained metallic antimony which contained higher component of gold is called gold-antimony alloy (Au-Sb alloy). Plants in china used to electrolyze the Sb-Au alloy directly for separation of Sb and Au. In electrolysis, Sb deposited on cathode while Au is remained in anode slime, then the subsequent recovery process for Sb and Au is simple. Nowadays, the impurities of Cu and Ni in mineral resource are becoming higher, the impurities will be enriched in Sb-Au alloy with Au, which caused the original electrolysis process hard to implement. The typical chemical components of Sb-Au alloy in 2000 and 2014 are given in Table Ⅰ.

Table Ⅰ. Chemical component of Sb-Au alloy (wt. %)

Year	Sb	Pb	Cu	Ni	Fe	As	Ag	Au
2000	~80	~5	1~5	<1	~1.5	~0.5	~0.5	>2.0
2014	~70	~15	>15	>3	~0.5	~0.5	~0.4	~1.5

In Table I, we can see that the impurities have great potential value, and it is important to comprehensive recovery of valuable metals in Au-Sb alloy. Yang [3] and Liu [4] treated with these Au-Sb alloy samples containing high impurities by selective chlorination leaching. The results showed that base metals (Cu, Ni, Sb) were removed completely from the alloy, and crude gold powder with grade of 94% were obtained. Sb, Cu and Ni in leaching solution were separated respectively by neutralization precipitation. It is obviously that the precipitate of Sb can entrain part of the Cu and Ni in the residue. The separation of Sb, Cu, and Ni is not thorough, which can cause mazy further processing of antimony products.

It is convenient for further processing of antimony products, if antimony can be separated completely with copper and nickel in Au-Sb alloy. Cu and Ni can be dissolved in sulphuric acid solution, while Sb and Pb can not be. Based on these characteristics of the base metals in Au-Sb alloy, an idea of separating Cu and Ni priorly from the alloy was explored by Liu [5]. The Au-Sb alloy was leached in different concentration of sulfuric acid solution (1.2~3.0 mol/L) at 80℃ for 2h at atmospheric pressure with L/S 7.5:1 and H_2O_2 1.0 times of theoretical, the leaching ratio of Cu was below 38%, and the leaching ratio of Ni was above 93%. The result indicated that Cu can not be completely dissolved at atmospheric pressure in sulphuric acid solution.

In order to intensify the dissolution of Cu in sulphuric acid solution, pressure leaching process for Au-Sb alloy is appropriate. In this work, the combined process of pressure oxidation leaching in sulphuric acid system (POLS) followed by hydrochloric acid leaching (HAL) was employed to remove base matals from Au-Sb alloy. The effects of H_2SO_4 concentration, liquid-solid ratio (L/S), consumption of oxidizer, leaching time and leaching temperature in the step of POLS on the metal leaching ratios of POLS and subsequent HAL of the residue from POLS step were investigated.

Experimental

Materials

The Au-Sb alloy used in the experiments was produced in a reverberatory furnace from a smelting plant in China. This sample was ground to 100 mesh before experiments. And the chemical compositions of this Au-Sb alloy powder were given in Table II.

Table II. Chemical compositions of Au-Sb alloy

Compositions	Sb	Pb	Cu	Ni	Fe	Ag	Au
wt.%	57.02	8.20	17.37	4.96	0.079	0.327	2.48

In this work, the main chemical reagents of H_2SO_4, HCl and the oxidizer of H_2O_2 are of analytical grade.

Process

The combined process of pressure oxidation leaching in sulphuric acid solution (POLS) followed by hydrochloric acid leaching (HAL) was employed to remove base metals from Au-Sb alloy. The process flowsheet is shown in Figure 1.

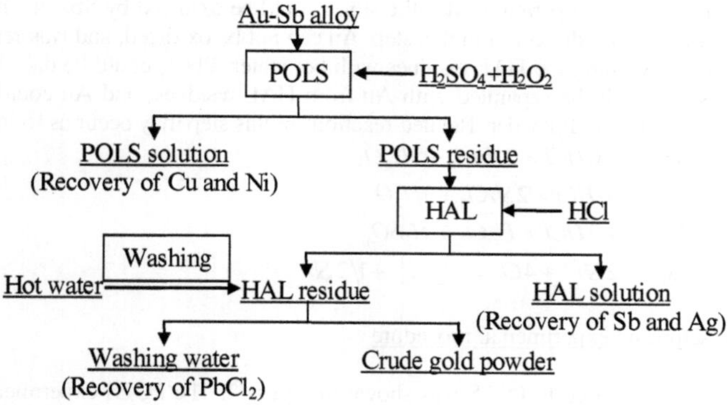

Figure 1. Flowsheet of Au-Sb alloy treatment

POLS In this step, Cu and Ni in Au-Sb alloy were completely oxidized and dissolved into POLS solution, leaving the oxides of Sb and Pb in the POLS residues which were then leached out using hydrochloric acid. Related reactions in this step may occur as follows

$$H_2O_2 = H_2O + O_2 \tag{1}$$
$$2Cu + O_2 + 2H_2SO_4 = 2CuSO_4 + 2H_2O \tag{2}$$
$$2Ni + O_2 + 2H_2SO_4 = 2NiSO_4 + 2H_2O \tag{3}$$
$$4Sb + 3O_2 = 2Sb_2O_3 \tag{4}$$
$$4Sb + 5O_2 = 2Sb_2O_5 \tag{5}$$
$$2Pb + O_2 + 2H_2SO_4 = 2PbSO_4 \downarrow + 2H_2O \tag{6}$$
$$4Ag + O_2 + 2H_2SO_4 = 2Ag_2SO_4 \downarrow + 2H_2O \tag{7}$$

In the autoclave, hydrogen peroxide releases oxygen gradually with the increase of temperature of the solution. In fact, oxygen is the oxidizer in the solution. The amount of H_2O_2 was represented as times of its theoretical consumption in this work. The theoretical consumption of H_2O_2 can be calculated from reaction (1)~(7) and the element component in Table II. In addition, gold can not be oxidized and dissolved, silver can be oxidized and partly dissolved in the POLS solution. The dissolution of Ag_2SO_4 in sulphuric acid solution was at least 8.5 mg/L by theoretical calculation under the condition in this step. In experiments, the concentration of Ag^+ in solution was analyzed with AAS. But there is no Ag_2SO_4 in POLS solution, which indicated that reaction (7) was not occurred in this step. Low content of Ag in Au-Sb alloy and insoluble compounds in POLS can hinder the reaction of metallic silver with oxidizer.

HAL In this step, residues from POLS step was treated directly without drying in hydrochloric acid solution. Oxides of Sb and Pb in POLS residues were dissolved into HAL solution in this step. Lead in HAL solution was formed as $PbCl_2$ which could be separated with antimony by crystallization at low temperature. Metallic silver could be oxidized by Sb^{5+} in chloride solution. Thus, Ag can be partly dissolved in this step. Au can not be oxidized, and was remained in HAL residues. While washing the HAL residues with hot water, $PbCl_2$ could be dissolved in washing water. Thus, Pb could be separated with Au from HAL residues, and Au could have a further enrichment in crude gold powder. Related reactions in this step may occur as follows

$$Sb_2O_5 + 10HCl = 2SbCl_5 + 5H_2O \quad (8)$$

$$Sb_2O_3 + 6HCl = 2SbCl_3 + 3H_2O \quad (9)$$

$$PbSO_4 + 2HCl = PbCl_2 + H_2SO_4 \quad (10)$$

$$Ag + 1/2\,Sb^{5+} + 4Cl^- = AgCl_4^{3-} + 1/2\,Sb^{3+} \quad (11)$$

Autoclave setup and experimental procedure

The autoclave setup used in POLS was shown in Figure 2. The POLS experiments were carried out in a 100ml stainless steel autoclave with a polytetrafluorethylene (PTFE) lining. The autoclave was heated in a sand bath. And a rotor was equipped for agitating. The rotor was rotated by a magnetic stirrer.

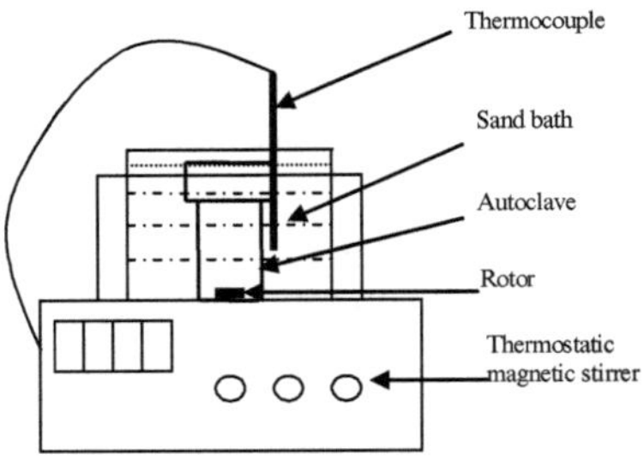

Figure 2. Autoclave setup in POLS

In POLS step, firstly H_2SO_4 solution, H_2O_2 and alloy powder (8 g per time) were added into the autoclave. Then the sealed autoclave was heated to the required temperature in sand bath. After reaction, the resultant slurry was vacuum-filtered till the autoclave was cooled to room temperature. The POLS residue was washed with de-ionized water for twice. After washing, the filter paper and wet residue were sent to next step directly.

In HAL step, residues from POLS were leached in 80ml hydrochloric acid solution with the concentration of HCl 5.0 mol/L at 80℃ for 1 hour. After reaction, the resultant slurry was vacuum-filtered till the solution was cooled to room temperature. The HAL residue was washed with hot water for twice and then dried in air oven at 60℃.

Analysis

The contents of Cu, Ni and Sb in solution were detected with the titration method of sodium thiosulfate, EDTA and cerium sulfate, respectively. The content of Ag in solution was detected with AAS for its low concentration in the solution.

Results and discussion

The variables of in POLS step not only had influence on POLS, but also affect the subsequent HAL step of POLS residue leaching. So it is necessary to investigate the effect of main variables of POLS on them. Such as the consumption of H_2O_2, the concentration of H_2SO_4, liquid-solid ratio (L/S), leaching time and leaching temperature.

Effect of the consumption of hydrogen peroxide

Effect of the consumption of H_2O_2 on POLS The effect of H_2O_2 consumption from 0.9 to 1.3 times of the theoretical consumption on leaching ratios was researched under the conditions of concentration of H_2SO_4 1.5 mol/L, L/S of 7:1, temperature 150 ℃ and leaching time 4 h. The results are shown in Figure 3.

It can be seen from Figure 3 that the consumption of H_2O_2 had a strong influence on the leaching ratio of Cu. That because Cu is the electropositive metal which can not be dissolved in acidic solution without oxidizer. The leaching ratio of Cu increased sharply till H_2O_2 consumption was 1.2 times of theoretical. Leaching ratio of Cu was above 99.0% when the consumption of H_2O_2 was over 1.2 times of theoretical.

Results in Figure 3 showed that the consumption of H_2O_2 had no influence on the leaching ratio of Ni. That because Ni is the electronegative metal which can be dissolved easily in acidic solution without any oxidizer. And the leaching ratios of Ni were all 100% in Figure 3. In fact, the leaching ratios of Cu were detected as 100~110% in POLS experiments under the research conditions except for leaching time of 2h. It means that Ni the leaching ratio of Ni was only affected by leaching time. It is obviously that there was a little error in analysis Ni or in sampling of the alloy materials. The sampling error was more credible than the analysis error. Because the scale in this work is only 8 g alloy for one test, and the content of Ni in the alloy is low. Subtle addition of the alloy can cause the obvious changes of the leaching ratio of Ni. Meanwhile, Ni in the alloy could be easily dissolved in POLS [5]. Thus, the results of Ni plotted as 100% in figures in this work are credible.

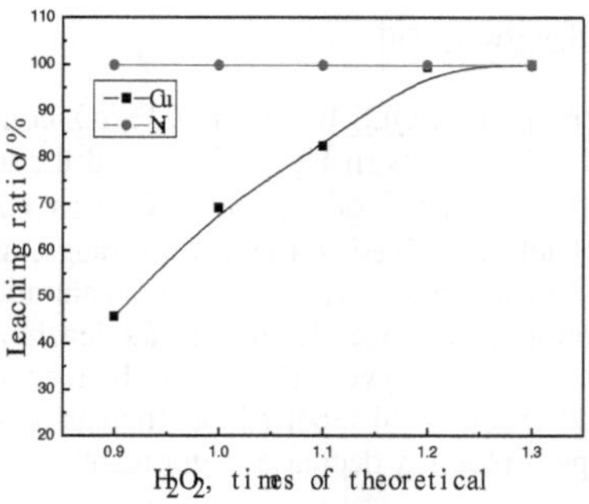

Figure 3. Effect of the consumption of hydrogen peroxide on POLS

Effect of the consumption of H_2O_2 on HAL The oxidizer consumption also had an indirect influence on the subsequent HAL of POLS residue. The results are listed in Figure 4.

Figure 4 showed that the leaching ratios of Sb increased slowly and it was up to 92.8% till the H_2O_2 consumption was 1.2 times of theoretical. When the consumption of H_2O_2 were above 1.2 times of theoretical, the leaching ratios of Sb decreased. As the increase of H_2O_2, the degree of oxidation of Sb improved. And the leaching ratios of Sb increased. But excessively addition of H_2O_2 could caused large amount of Sb_2O_5 in POLS residue. Sb_2O_5 is laboriously dissolved in hydrochloric acid solution. And the leaching ratios of Sb decreased with high consumption of H_2O_2. In fact, titration method for Sb analysis [6] indicated that Sb^{5+} was always existed in HAL solution even the consumption of H_2O_2 was 0.9 times of theoretical. Thus, the oxidization and dissolution Ag in HAL were avoidless. Figure 4 showed that leaching ratios of Ag increased quickly with the increase of H_2O_2. According to the results of above, H_2O_2 1.2 times of theoretical is chosen as the optimal condition.

In addition, the concentration of Pb was not analyzed in this work. Because Pb is the electronegative metal which can be oxidized easily into $PbSO_4$ during POLS step, and can be dissolved easily in hot solution as $PbCl_2$ during HAL step. But $PbCl_2$ was removed from solution for separation with Sb by crystallization at room temperature, and then was washed into washing water.

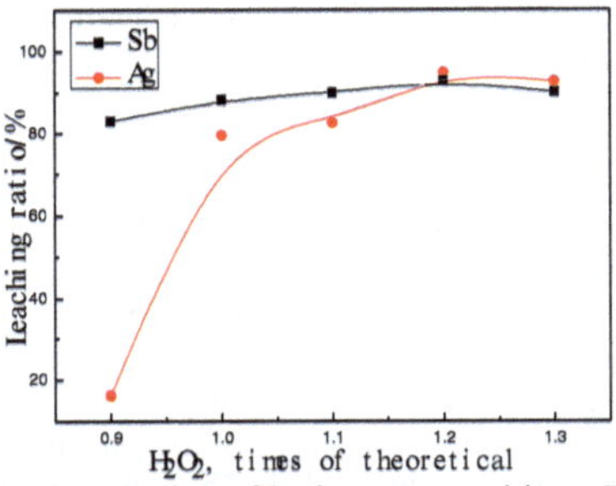

Figure 4. Effect of the consumption of hydrogen peroxide on HAL of POLS residue

Effect of the concentration of sulfuric acid

Effect of the H_2SO_4 concentration on POLS In this step, H_2SO_4 concentration was studied in the range from 0.5 to 2.0 mol/L. The results in Figure 5 showed the leaching ratio of metals from alloy leached for 4 h at 150 ℃ with L/S of 7:1 and H_2O_2 was 1.2 times of theoretical. It can observed that Ni was completely dissolved in this acidity range, and the leaching ratios of Cu increases strongly from 45.0% to 99.5% with H_2SO_4 concentration increase from 0.5 to 1.5 mol/L. When H_2SO_4 concentration was over 1.5 mol/L, the leaching ratios of Cu decrease. The decrease could be due to a decrease of oxygen solubility in high concentration of H_2SO_4 solution. On this matter, Narita [7] and Tromans [8] reported that increasing sulfuric acid concentration in an aqueous solution the oxygen solubility decreases significantly.

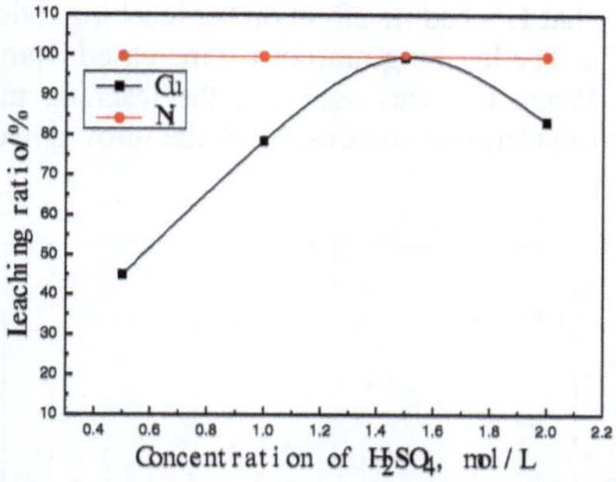

Figure 5. Effect of sulfuric acid concentration on POLS

Effect of the H_2SO_4 concentration on HAL Meanwhile, the H_2SO_4 concentration of POLS also had a significant influence on the subsequent HAL of POLS residue. The results are shown in Figure 6. It can be seen that the leaching ratios of Sb and Ag had the similar variation tendency with different H_2SO_4 concentrations. The leaching ratios of Sb and Ag decreased to 82.7% and 78.4% respectively when H_2SO_4 concentration was over 1.5 mol/L. Because that the leaching ratio of Ag was affected directly by the concentration of Sb^{5+} in hydrochloric acid solution. And the concentration of Sb^{5+} in hydrochloric acid solution was combined closely with the leaching ratio of total Sb. The degree of oxidation of Sb in POLS dominated the dissolution of antimony oxides in HAL. According to the results above, H_2SO_4 concentration of 1.5 mol/L was the optimal condition.

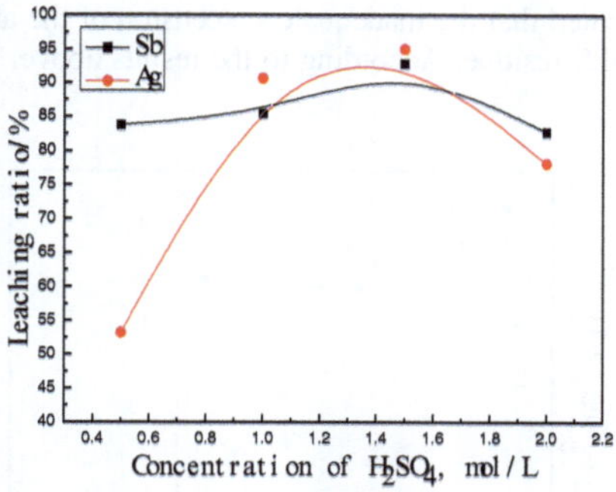

Figure 6. Effect of sulfuric acid concentration on HAL

Effect of L/S

Effect of L/S on POLS The effect of liquid-solid ratio was studied in the range 5:1 to 8:1. On the condition of 1.5 mol/L sulfuric acid and hydrogen peroxide 1.2 times of theoretical, the alloy is leached for 4 h at 150 ℃. In this step, the leaching ratios of metals were plotted in Figure 7.

It can be seen from Figure 7 that L/S had no effect on the leaching ratio of Ni, but had significant effect on leaching ratio of Cu. The leaching ratio of Cu increased from 67.9% to 99.5% with L/S increased from 5:1 to 7:1. When L/S was over 7:1, the leaching ratio of Cu decreased. The decrease could be due to an inadequate suspension of the alloy powder in high column of the leaching solution.

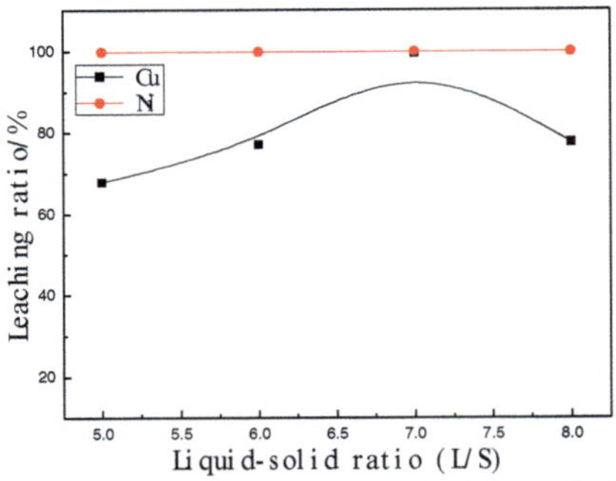

Figure 7. Effect of liquid-solid ratio on POLS

Effect of L/S on HAL Similarly, liquid-solid ratio of POLS had relevant effects on HAL. It can be seen from Figure 8 that the leaching ratios of Sb increased with low L/S, and changed smoothly with high L/S. The leaching ratios of Ag increased gradually with the increase of L/S. Which means the concentration of Sb^{5+} in hydrochloric acid solution was increased with the increase of L/S. And indicated that the inadequate suspension of the alloy could promote the the formation of Sb_2O_5 in POLS residue. According to the results above, L/S of 7:1 was considered the optimal.

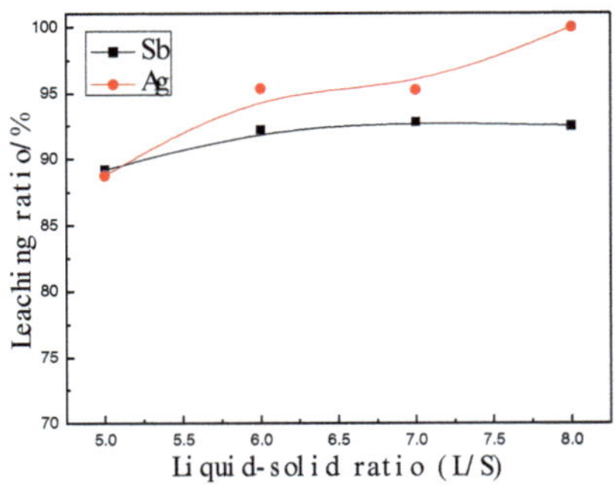

Figure 8. Effect of liquid-solid ratio on HAL

Effect of leaching time

Effect of leaching time on POLS The effect of leaching time was studied in the range 2 to 5 h. On the condition of 1.5 mol/L sulfuric acid and hydrogen peroxide 1.2 times of theoretical, the alloy is leached at 150 ℃ with L/S of 7:1. Leaching ratios of metals from alloy was shown in Figure 9.

The results in figure 9 showed that leaching time had effect on the dissolution of Ni, Ni can not be completely dissolved when the alloy is leached for less than 2 h. It can also be seen that leaching ratios of Cu increased strongly from 3.3% to 99.5% with the leaching time increased from 2 to 4 h. When leaching time was over 4 h, the leaching ratio of Cu decreased. The results indicated that Cu was completely dissolved in the solution when the alloy was leached for 4 h. After 4 h, Sb and Pb could be largely oxidized and forming insoluble compounds. These insoluble compounds suspended in the solution and absorbed Cu^{2+} from the solution that caused the leaching ratio of Cu decreased. Thus, leaching time should be controlled as 4 h.

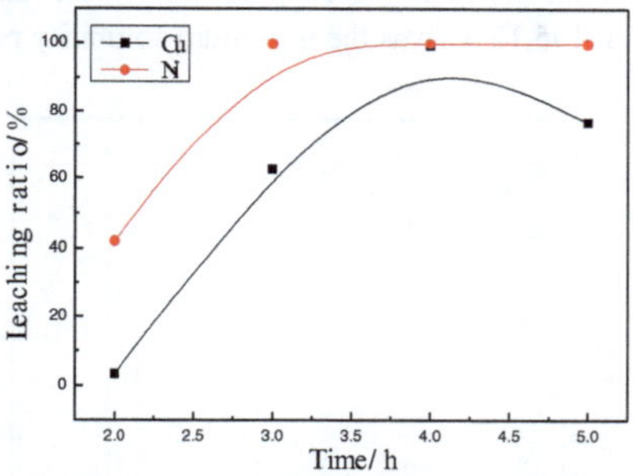

Figure 9. Effect of leaching time on POLS

Effect of leaching time on HAL It is obviously that enhance the reaction time can cause more and more Sb_2O_3 changing into Sb_2O_5 in POLS. That could cause the relevant variations in HAL. The results were shown in Figure 10. It can be seen from Figure 10 that leaching ratios of Sb increased from 36.6% to 92.8% with the leaching time increased from 2 to 4 h. When leaching time was over 4 h, the leaching ratio of Sb decreased. That indicated that more occupation of Sb_2O_5 in total amount of Sb in POLS residue caused this decrease. It can also be seen that leaching ratios of Ag were increased gradually with the increase of leaching time. According to the results above, leaching time of 4h was adequate.

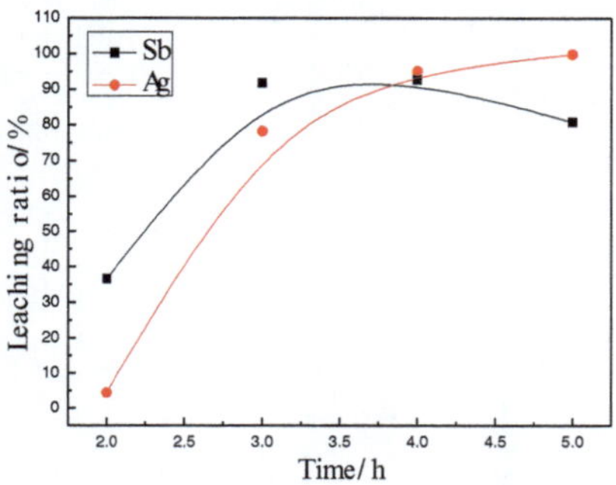

Figure 10. Effect of leaching time on HAL

Effect of leaching temperature

Effect of leaching temperature on POLS The effect of leaching temperature was studied in the range 120 to 180 ℃. On the condition of 1.5 mol/L sulfuric acid and hydrogen peroxide 1.2 times of theoretical, the alloy was leached for 4 h at 150 ℃ when L/S 7:1. Leaching ratios of metals from alloy was plotted in Figure 11. It can be seen in Figure 11 that leaching temperature had no effect on the leaching ratio of Ni. The leaching ratio of Cu increased from 66.4% to 99.9% at temperature increased from 120 ℃ to 150℃, the leaching ratio of Cu decreased when temperature was over 150℃. The decrease could be due to a decrease of oxygen solubility in this temperature range. On this matter, Tromans [9] reported that increasing temperature in the range T< 368.12 K in sulfuric acid solution (or in pure water) the oxygen solubility decreased significantly. Temperature at 195.12 ℃ was the minimum solubility point of oxygen.

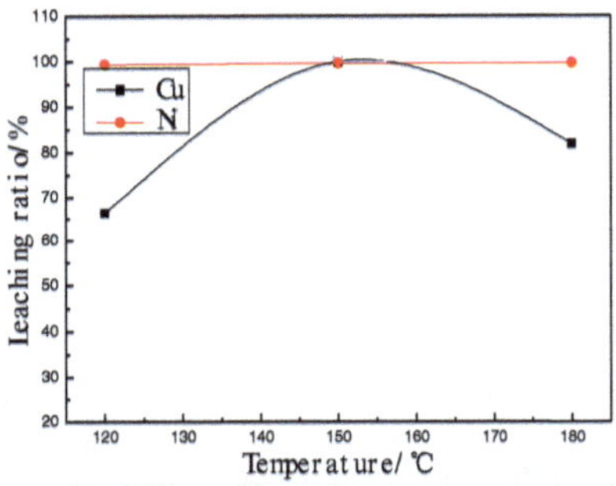

Figure 11. Effect of leaching temperature on POLS

Effect of leaching temperature on HAL Similarly, the solubility rule of oxygen in POLS had influence on the oxidation degree of Sb in the POLS residue. That could affect the dissolution of Sb and Ag in HAL. It can be seen from Figure 12 that the leaching ratio of Sb decreased with the solubility of O_2 decrease at high temperature. But the leaching ratio of Ag always increased at the research temperatures. That could be due to the solubility of AgCl increased at high temperature. According to the results above, 150℃ was chosen as they optimal temperature.

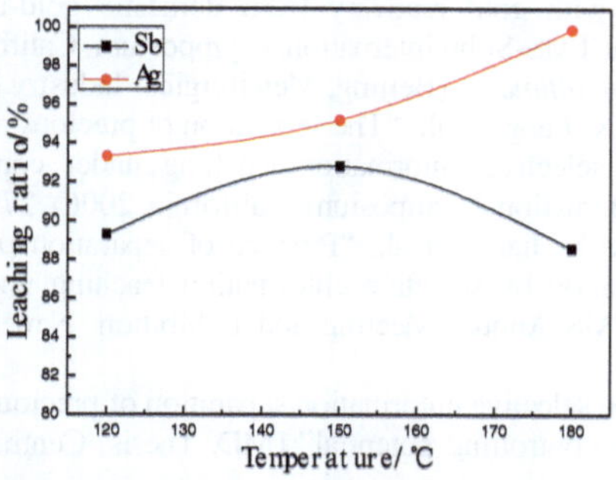

Figure 12. Effect of leaching temperature on HAL

Optimum conditions

In order to demonstrate the optimal conditions got from single factors experiments, two replicated experiments were carried out on the optimal conditions. The alloy (8 g) was leached in1.5 mol/L H_2SO_4 at 150 ℃ for 4 h when H_2O_2 1.2 times of theoretical and L/S 7:1. The POLS residues were then treated in HAL as conditions of HCl 5.0 mol/L, temperature of 80℃ and time 1 h. The results were listed in Table III.

Table III. Results of confirmation experiments under the optimum conditions

	No.	Cu	Ni	Sb	Ag	Grade of crude gold powder/%
POLS/%	1	99.5	100	-	-	-
	2	99.9	100	-	-	-
HAL/%	1-1	-	-	93.5	95.6	66.8
	1-2	-	-	92.3	94.2	66.1

Conclusions

① Cu, Ni can be priorly separated from Au-Sb alloy by pressure oxidation leaching.
② Base metals Cu, Ni, Sb and Pb can be removed fully for enrichment of gold.
③ The leaching ratio of Ni is only be affected by the leaching time.

Acknowledgements

The author appreciates professor YANG Tianzu for his help in this work. The project was supported by the Natural Science Foundation of Jiangsu Province (BK20140558) and the Advanced Foundation of Jiangsu University (13JDG097).

References

1. F.A. Myzenkov, A.V. Taraov, "Development and creation of modern metallurgical technology

for antimony production with gold recovery from domestic gold-antimony concentrates in Russia" (Paper presented at TMS-Sohn international symposium, California, 2006), 121-128.
2. T.C. Zhao, *Metallurgy of antimony*, (Beijing, Metallurgical Industry Press, 1987), 495-502.
3. T.Z. Yang, W.F. Liu, M.X. Jiang, et al., "The separation of precious metals from base metals in gold-antimony alloys by selective chlorination leaching under controlled potential" (Paper presented at TMS-Sohn international symposium, California, 2006), 279-292.
4. W.F. Liu, T.Z. Yang, M.X. Jiang, et al., "Practice of separation precious metals from base metals in gold-antimony alloy by selective chlorination leaching under controlling potential" (Paper presented at the TMS Annual Meeting and Exhibition, New Orleans LA, 9-13 March 2008), 479-485.
5. W.F. Liu, "Studies on the selective chlorination separation of precious metals from base metals in gold-antimony alloy by controlling potential" (MD. Thesis, Central south university, 2006), 30-31.
6. Y.Y. Gu, D. Li, "Determination of antimony (III), antimony (V) and hydrogen peroxide in the oxidized antimony white solution by hydrogen peroxide," *Metallurgical Analysis*, 4(2006), 20-23.
7. E. Narita, F. Lawson, K.N. Han, "Solubility of oxygen in aqueous electrolyte solution," *Hydrometallurgy*, 10 (1983), 21-37.
8. D. Tromans, "Temperature and pressure dependent solubility of oxygen in water: a thermodynamic analysis," *Hydrometallurgy*, 48 (1998), 327-342.
9. D. Tromans, "Oxygen solubility modeling in inorganic solutions: concentration, temperature and pressure effects," *Hydrometallurgy*, 50 (1998), 279-296.

Rare Metal Technology 2016
Edited by: Shafiq Alam, Hojong Kim, Neale R. Neelameggham, Takanari Ouchi, and Harald Oosterhof
TMS (The Minerals, Metals & Materials Society), 2016

QUANTITATIVE ANALYSIS OF THE TRACE ELEMENTS IN PURITY INDIUM MATERIAL BY GLOW DISCHARGE MASS SPECTROMETER

Wenli Zhang[1,2], Ping Long[1,2,*], Jian Wu[1,2], Xiumin Chen[1,2], Bin Yang[1,2]

[1]National Engineering Laboratory for Vacuum, Kunming University of Science and Technology, Kunming 650093, Yunnan, China
[2]Key Laboratory for Nonferrous Vacuum Metallurgy of Yunnan Province, Kunming 650093, Yunnan, China)

Keywords: Glow discharge mass spectrometry, Pure indium, Trace element quantitative analysis.

Abstract

In this study, the analysis method of trace element in pure indium using glow discharge mass spectrometry (GD-MS) has been established by correcting the analysis result, which was obtained from non-standard quantitative method of GD-MS. The effect of instrument operating parameters, the resolution and isotopes and the pre-sputtering time on the determination results were investigated respectively. The optimized operating parameters of the instrument were obtained as follows: the discharge current was 37.5mA and discharge airflow was 475ml/min. Mass spectral interferences have been eliminated using selected resolution and isotope. The pre-sputtering time was 30 min. 4 trace elements in two pure samples were determined for 6 times. The RSD were less than 20%. The content of trace elements in indium samples was determined by GD-MS and Chemical Analysis Method of Indium which is tested by inductively coupled plasma atomic emission spectrometry, respectively. The results were consistent with each other.

Introduction

Pure indium has been used to fabricate ITO (indium tin oxide) target and is also widely used in the field of semiconductor, solder and alloys [1-2] because of its plasticity, low melting point, high boiling point, low resistance, corrosion resistance and other excellent properties. Impurities in pure indium affect the application of pure indium in high-tech fields. Therefore, the research on fast and accurate determination of trace element in pure indium has attached a lot of attention. Currently, the methods, such as chelatometric titration, atomic absorption spectroscopy and spectrophotometry [3-4], are applied to determine pure indium generally. But the process of these methods is slow and complex.

Glow discharge mass spectrometry (GD-MS) has been widely applied in electronics, chemistry, metallurgy and materials science for its special advantages, including analyzing solid samples directly, determining multi-element simultaneously and rapidly, high-resolution in quality, accurate in result and great precision [5-12].It was difficult to obtain solid standard sample (I have explained it in the end of the paper) so that non-standard quantitative method is used in GD-MS determination generally. Liu

* Corresponding author: Tel/fax: 871-65161583
E-mail: pkulongping@sohu.com (P. Long), 1499608903@qq.com (W. L. Zhang).

et al [13] have measured the iron, copper, aluminum, zinc and 14 kinds of elements in high-purity indium by the non-standard quantitative method of GD-MS. However, their studies have not met the requirements for accurate quantification of some elements in pure indium. In this study, GD-MS was used as the main research tool. The measuring method, which determines four trace elements (cadmium, tin, lead, bismuth) in pure indium material simultaneously, was established by correcting the result of non-standard quantitative. The research results provide evidence for the determination of pure indium materials.

Experimental

Instruments and operating conditions

Measurements, such as glow discharge mass spectrometer (Thermo Fisher Scientific), OPTIMA 8000 inductively coupled plasma atomic emission spectrometry (PerkinElmer), WFX-320 Atomic Absorption Spectrophotometer (Beijing Rayleigh Analytical Instrument Co., Ltd) and 721 spectrophotometer (Shanghai precision &scientific instrument Co., Ltd), were used. Table 1 summarized operating conditions for the analyses of high purity samples for GD-MS.

Table 1 Operating conditions of GD-MS

Working parameters	Numerical	Working parameters	Numerical
Discharge Current/mA	37.5	Extraction /V	-2000
Discharge Voltage /V	365	Focus /V	-1078
Discharge Gas /(mL/min)	475	X-Deflection/V	0.69
Source Vacuum/mbar	1.07×100	Y-Deflection /V	1.51
Fore Vacuum/mbar	2.05×10-3	Shape/V	157
High Vacuum/mbar	2.91×10-7	Filter Lens /V	4

Experimental material and reagents

A series of high purity indium (5N) samples were used for the experiments. Reagents were nitric acid, hydrochloric acid, sulfuric acid, tartaric acid, potassium permanganate, and phenylfluorone and bromo cetyl trimethylamine. The seven reagents were guarantee reagent. Water was ultra-pure water with resistivity greater than 18MΩ·cm. A series of single element standard solutions of cadmium, tin, lead and bismuth with concentration of 1 mg/mL were purchased from National Research Center.

Experimental methods

Samples were prepared as follows: the raw materials were manufactured into lots of sheet samples with the size of 20mm× 20 mm and thickness of 5 mm. These samples were cleaned with 20% dilute nitric acid in ultrasonic condition and repeatedly washed more than three times using ultrapure water. Then they were dehydrated by anhydrous ethanol and taken out. These samples were dried free and then tested.
Samples were pre-sputtered as follow: the samples were put into the sample holder and were pushed into ion source cavity. Discharge current, discharge voltage and gas flow were 37.5mA, 365V and 475 ml / min respectively. Sputtering time was

approximately 30 min. Generally, in order to avoid the possible pollution of sample in preparation process, the stability of Na and Ca signal were considered mainly for their characteristic of easy introducing impurities. (I have explained it in the end of the paper)

Samples were measured as follow: high voltage discharge was proceeding when the discharge voltage was controlled at 365V and the discharge current was controlled at 37.5mA. Then, ion signals of determined elements were collected. The instrument automatically showed the none-standard quantitative results of impurity elements.

The results of GD-MS were not accurate, so we calculated the results to obtain the quantitative analysis. Measurement results were revised as follow: the results of GD-MS of cadmium, tin, lead and bismuth were calculated as following equations (I have explained it in the end of the paper) from (1) to (4) (where x is measured value of non-standard quantitation): $y=0.6005x-0.1143$ (1), $y=0.7588x-2.0756$ (2), $y=0.7032x-1.9890$ (3),
$y=0.7348x-27.7909$(4).

Results and discussion

Optimization of the main operating parameters

According to the discharge current which changed in the range of 15-50mA and the discharge gas flow which changed in the range of 375-500mL / min, the change of indium signal strength value was investigated, as shown in figure 1 and 2. The best parameters (discharge current and discharge gas flow) was obtained when the signal was stable, the signal strength was greater than 1×10^{10}CPS (counting per second) at low resolution and instrument resolution reached 300.

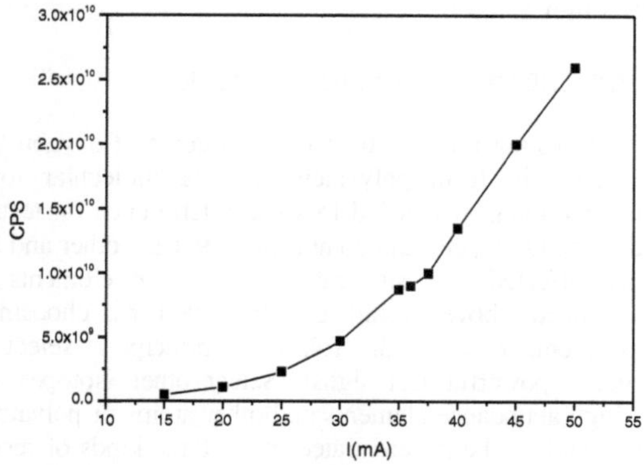

Figure.1 Effect of discharge current (I) on CPS of GD-MS

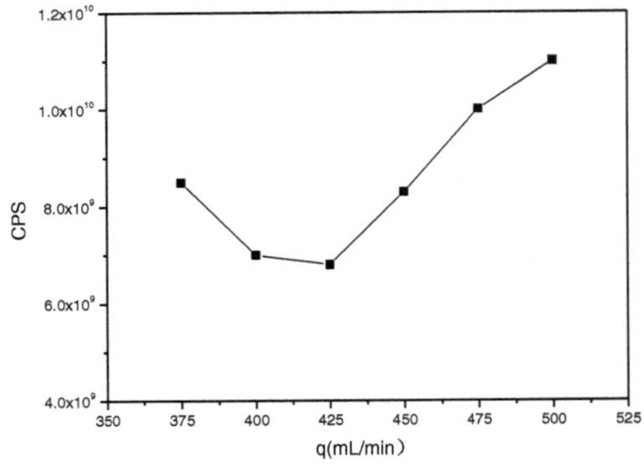

Figure.2 Effect of discharge gas flow (q) on CPS of GD-MS

It can be seen from Figure.1 that the indium matrix signals become stronger and stronger when the discharge current changes from 15mA to 50mA. It can be also noted from Fig.1 that CPS value can reach 1×10^{10} and resolution met requirements when the current was greater than 37.5mA. However, if the current was too high, the spilled indium quickly attached to the anode cap and sample temperature became too high, which led to a short circuit. Therefore, the discharge current had be chosen at 37.5mA in this experiment.

Figure.2 suggested that CPS value first decreased and then increased when the gas flow rate varied in the range from 375 mL / min to 500mL / min. When the gas flow was greater than 475mL / min, CPS value can reach 1×10^{10}. At the same time, signal was stable and resolution met requirements. Thus the discharge gas flow rate was chosen at 475mL / min.

Selections of isotopes and resolution of the instrument

During the GD-MS testing process, high purity argon and oxygen, carbon, nitrogen and other elements will form polyvalent ions or molecular ions for multiple discharges or polymerizing, which led to some interference on testing elements. In addition, In and H, N, O, Ar, etc. can composite with each other and form polyatomic ions, which also affected analysis and test of some elements. Therefore, the interference mentioned above would be eliminated by choosing isotopes and resolution. Experiments base on the following principles: select higher isotopic abundance to get a powerful test signal. Select other isotopes which could be distinguished if high abundance elements and other atoms or polyatomic ions caused interference and could not be differentiated under three kinds of resolutions. Isotope and resolution of the instrument could be optimized as these principles descriped above. The results are showed in Table 2.

Table 2 Optimization of isotopes and resolution of the instrument

Element	Cd	Pb	Sn	Bi
Isotope	^{114}Cd	^{206}Pb	^{118}Sn	^{209}Bi
Resolution	Intermediate	Intermediate	Intermediate	Intermediate
Interference	nothing	nothing	nothing	nothing

Selection of pre-sputtering time

Because the sample surface will be contaminated inevitably in the preservation and pre-treatment process, pre-sputtering of sample was necessary in order to ensure the accuracy and stability of impurity elements signal. Generally, pre-sputtering could consider to be stopped when the elements signal, which was vulnerable to pollution, such as Na, Ca, would not fluctuate with the increase of sputtering times. Figure 3 shows the graph about the contents of impurity elements Na, Ca with the variation of time. The contents of Na and Ca decrease in the range of 2.5~30 min because pre-sputtering of sample strip the part of Na and Ca as pollutants on the sample surface. It will not change after 30 min because Na and Ca pollutants have been stripped completely. Pre-sputtering time could be selected at 30 min, as shown in Figure 3.

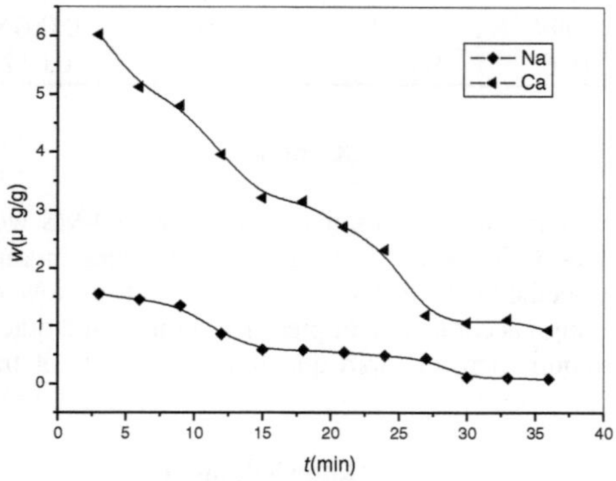

Figure.3 Curve about contents of Na and Ca with sputtering time

Corrected linear regression equation and the correlation coefficient

Four impurity elements in five samples of pure indium were determined by the chemical analysis of indium (YS/T276-2011) and non-standard quantitative method of GD-MS respectively. The linear regression equations and the correlation coefficient were obtained by linear fitting of measurement result (where x is no-standard quantitative value and y is corrected value), as showed in Table 3. Linear range of cadmium, tin, lead and bismuth were 0.2~25μg/g, 700~25000μg/g, 40~400μg/g and 100~2000μg/g, respectively.

Table 3 Linear equations and the correlation coefficient of four metals

Element	The regression equation	The correlation coefficient
Cd	y=0.6005x-0.1143	0.9977
Sn	y=0.7588x-2.0756	0.9999
Pb	y=0.7032x-1.9890	0.9995
Bi	y=0.7348x-27.7909	0.9994

The precision and accuracy of the method

Two samples were measured in parallel for six times in order to verify the precision and accuracy of the method. The comparisons of the results obtained above with that of chemical analysis method are shown in Table 4. Experimental results indicated that the relative standard deviation of them were in the range of 1.0%~17.5%. the results determined by GD-MS and chemical analysis method of indium were consistent with each other.

Table 4 Sample analysis and results of precision/%

Element	The measured values of indium by GD-MS		The measured values of indium by chemical analysis		RSD	
	Sample1	Sample2	Sample1	Sample2	Sample1	Sample2
Cd	0.00017	0.000016	0.00016	0.000019	8.6	17.5
Sn	0.52	2.35	0.52	2.36	1	1.3
Pb	0.0099	0.008	0.0106	0.0078	2.1	2.3
Bi	0.036	0.012	0.033	0.012	1.7	1.6

Conclusion

The results of non-standard quantitative method of GD-MS were corrected and the analysis method of GD-MS for trace elements in pure indium material has been established. The method had many advantages, such as low detection limit, wide measuring range, high accuracy, great precision, simple and quick in operation, which could meet the requirement of rapid quantitative analysis of trace elements in pure indium material.

Acknowledgments

The research leading to these results was supported by the National Basic Research Program of China (No.2012CB722803).

References

[1] J.M. Zhong, K. Yue, B.S. Sun, D.X. Wang, "Fabrication process of large size ITO target materials by slip casting," The Chinese Journal of Nonferrous Metals, 09(2014), 2333-2338.

[2] H.M. Zhang, R. Chen, "Determination of Intium in ITO Powder and ITO Target by EDTA Titration," Chinese Journal of Rare Metals, 1(27)(2003), 187-189.

[3] YS/T276-2011, Methods for chemical analysis of indium.

[4] J. Cheng, X.L. Liu, "Direct determination of thallium in refined indium by flame atomic absorption spectrometry," Metallurgical Analysis, 06(2004), 66-68.

[5] R. Matschat, J. Hinrichs and H. Kipphardt, "Application of glow discharge mass spectrometry to multielement ultra-trace determination in ultrahigh-purity copper and iron: a calibration approach achieving quantification and traceability," Analytical And Bioanalytical Chemistry, 386(1) (2006), 125-141.

[6] T. Gusarova, B. Methven and H. Kipphardt. "Calibration of double focusing Glow Discharge Mass Spectrometry instruments with pin-shaped synthetic standards," Spectrochimica Acta Part B, 66(2011), 847-854.

[7] J.L. Dong, R. Qian and W. Xiong, "Determination of doping elements of synthetic crystals by direct current glow discharge mass spectrometry," International Journal of Mass Spectrometry, 361(2014), 1-8.

[8] R. Qian, B.L. Siqin, S.J. Zhuo, R.X. Shen, C. Sheng and F.X. Gan, "Analysis of Particle Hafnium by Direct Current Glow Discharge Mass Spectrometry," Chinese Journal Analytical Chemistry, (5)(2011), 700-704.

[9] M.Di Sabatino, "Detection limits for glow discharge mass spectrometry (GD-MS) analyses of impurities in solar cell silicon," Measurement, 50(2014), 135-140.

[10]C.Modanese, L.Arnberg and M.Di Sabatino, "Analysis of impurities with inhomogeneous distribution in multicrystalline solar cell silicon by glow discharge mass spectrometry," Materials Science and Engineering B, 180(2014), 27-32.

[11]J.W.Lim, K.Mimura and M.Isshiki, "Application of glow discharge mass spectrometry for direct trace impurity analysis in Cu films," Applied Surface Science, 27(2004), 300-305.

[12]V.Hoffmann, M.Kasik, PK.Robinson and C.Venzago, "Glow discharge mass spectrometry Anal," Bioanal. Chem., 381(2005), 173-188.

[13] H.Liu, A.C.Li and J.D.Li, "Determination of trace elements in high purity In by glow discharge mass spectrometry," Chinese Journal of Analysis Laboratory, 31(2012), 10-12.

Rare Metal Technology 2016

Platinum Group Metals / Mo, Ti, V & W

Rare Metal Technology 2016
Edited by: Shafiq Alam, Hojong Kim, Neale R. Neelameggham, Takanari Ouchi, and Harald Oosterhof
TMS (The Minerals, Metals & Materials Society), 2016

ADSORPTIVE RECOVERY OF PALLADIUM AND PLATINUM FROM ACIDIC CHLORIDE MEDIA USING CHEMICALLY MODIFIED PERSIMMON TANNIN

Manju Gurung[1], Birendra Babu Adhikari[1], Katsutoshi Inoue[1], Hidetaka Kawakita[1], Keisuke Ohto[1], Shafiq Alam[2]

[1]Department of Applied Chemistry, Faculty of Science and Engineering, Saga University
Honjo-machi 1, Saga 840-8502, Japan
[2]Department of Chemical and Biological Engineering, College of Engineering, University of Saskatchewan
Saskatoon, SK, S7N 5A9, Canada

Keywords: Persimmon tannin, Chemical modification, Adsorption, Palladium, Platinum, Acidic chloride media

Abstract

Novel bio-adsorbent gel was prepared by immobilizing functional groups of glycidyltrimethyl ammonium chloride, a kind of quaternary ammonium chloride, onto polymer matrices of persimmon extract for the recovery of precious metals such as palladium(II) and platinum(IV) from chloride media. It was found that this adsorption gel exhibits high selectivity to gold(III), palladium(II) and platinum(IV) over base metals such as copper(II) and iron(III) while crude persimmon extract, the feed material, exhibits high selectivity only for gold(III). From the adsorption isotherms study, the maximum adsorption capacities for gold(III), palladium(II) and platinum(IV) were evaluated as 3.30, 1.67 and 1.00 mol/kg, respectively. Those for palladium(II) and platinum(IV) are higher than those of simple quaternary ammonium chloride type of chemically modified persimmon tannin gel and commercially available anion exchange resin.

Introduction

Within the last decades, increasing attentions have been paid towards energy saving hydrometallurgical process for the recovery of valuable metals, precious metals in particular. In hydrometallurgical processing, a leach liquor is subjected to metal separation and purification, which includes solvent extraction, ion exchange, precipitation, membrane filtration and so forth. However, all the aforementioned methods suffer from various drawbacks, which include incomplete metal recovery, high capital costs, high reagent and/or energy consumption and generation of toxic sludge or other waste products that demand additional disposal considerations, which result in the increase of the overall cost. Due to their high efficiency, flexibility, applicability over a wide range of metal concentration, and simple sludge free operation for the effective recovery of analyte, adsorption techniques demonstrate great promise for the recovery of metal values from aquatic environment. From technical and economical points of view, in recent years, much attention has been focused on low cost adsorbents prepared from various biomass wastes as they are relatively inexpensive, environmentally benign, biodegradable and renewable. In such biomass wastes, the chemical components which exhibit effective adsorption behaviors include several polysaccharides such as pectic acid, alginic acid, chitin and cellulose, polyphenol compounds such as lignin and tannin, and proteins. Additionally, such adsorption behaviors of the original components can be significantly enhanced by means of chemical modification processes.

It is well known that tannin compounds, kinds of polymeric polyphenol compounds, exhibit ion exchange and complexation characteristics towards metal ions; that is, the potential of tannin rich biomass materials for the recovery of heavy and precious metals has been reported by some authors [1,2]. Persimmon is a popular fruit in East Asian countries such as China, Korea and Japan. There are two kinds of persimmon fruits, sweet persimmon and astringent persimmon, which differ by their content of water soluble persimmon tannin. Here, persimmon tannin is also a kind of polyphenol compound possessing a complicated chemical structure as shown in Scheme 1 containing large amounts of catechol and pyrogallol functional groups [3].

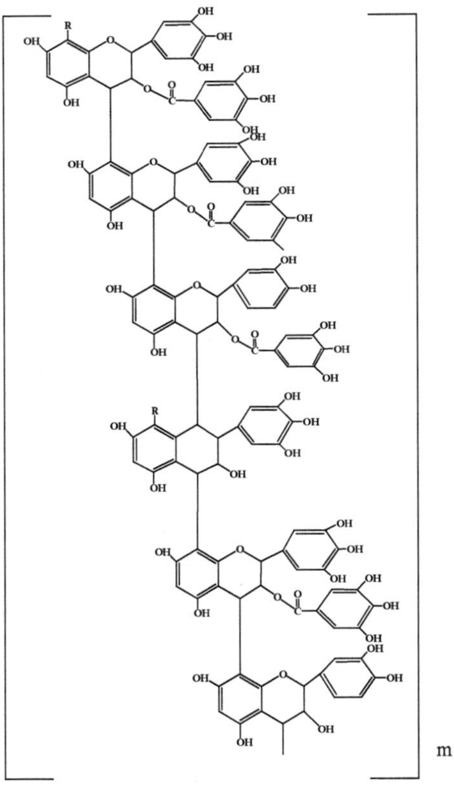

Scheme 1 Schematic structure of persimmon tannin where m stands for the number of units.

Persimmon extract, the juice of astringent persimmon, contains large quantities of persimmon tannin and has been traditionally employed in various applications as natural paints, dyes, tanning agent for leather tanning, a coagulating agent for proteins, etc. During the preparation of dried persimmon fruits and astringent persimmon extract, large quantities of peels and juice residue, or biomass wastes, are generated. These wastes still contain large amounts of persimmon tannin. In our recent works, environmentally benign adsorption gels were prepared from the powder of the astringent persimmon juice and also from the above-mentioned persimmon wastes by means of a simple and, consequently, inexpensive method, only by treating in boiling concentrated sulfuric acid as potential agents for the selective recovery of gold(III) [4]. This gel is abbreviated as CPT (Cross-linked Persimmon Tannin) gel, hereafter.

Figure 1 shows the results of the batch-wise adsorption tests as the plots of % adsorption of various metal ions from hydrochloric acid solution on CPT gel vs. hydrochloric acid concentration. As seen from this figure, quantitative adsorption of gold(III) was achieved on CPT gel over the whole concentration range of hydrochloric acid. Although this adsorbent exhibited about 10-20% adsorption for platinum(IV) and palladium(II), the adsorption of base metals was practically negligible under the present experimental conditions.

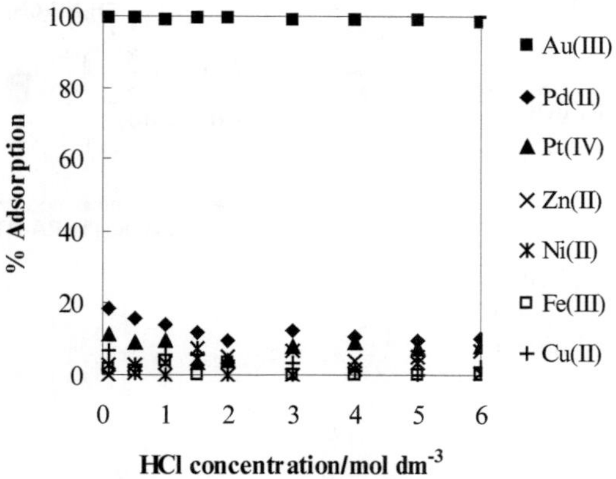

Figure 1. % adsorption of various metal ions on CPT gel as a function of concentration of hydrochloric acid. Conditions: initial metal concentration = 0.2 mmol dm^{-3}, weight of powder = 10 mg, volume of test solution = 10 ml, shaking time = 48 h, temperature = 303 K. [4]

Similar adsorption behaviors were observed also for persimmon tannin extract powder, the feed material of CPT gel [4] and for the gel prepared from persimmon peel generated in the production of dried persimmon fruits [5]. The remarkably high selectivity of CPT gel as well as the persimmon tannin extract powder for gold(III) over base metals as well as platinum(IV) and palladium(II) in a wide concentration range of hydrochloric acid is immensely useful for selective and quantitative recovery of gold(III) from acidic chloride media. It was inferred that the extraordinary high selectivity for gold(III) over other metal ions is attributable to the special reduction reaction catalytically performed by the molecules of cellulose cross-linked by concentrated sulfuric acid as well as also cross-linked persimmon tannin.

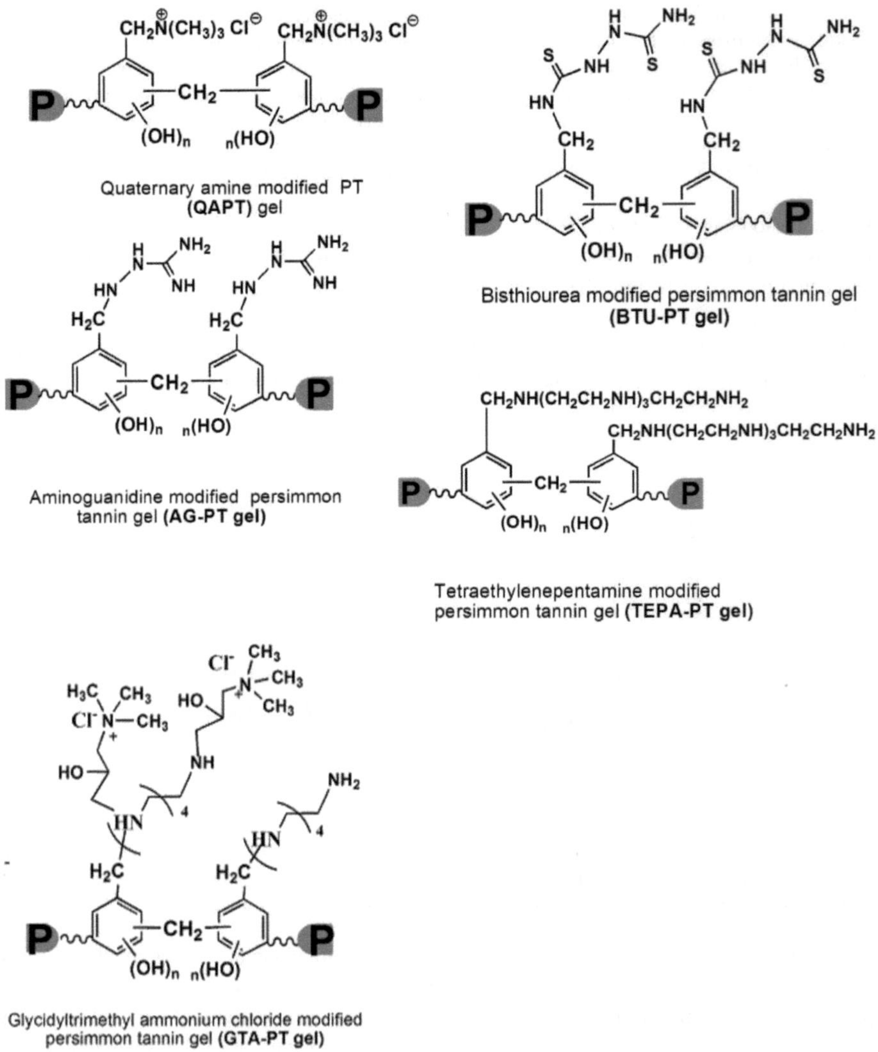

Scheme 2 Chemical structures of the chemically modified persimmon tannin gels prepared by the authors.

That is, although not only the powder of persimmon tannin extract, the feed material, but also the gels prepared from such powder and persimmon peel in boiling concentrated sulfuric acid exhibited extraordinary high affinity only for gold(III), only weak adsorption was observed for platinum(IV) and palladium(II). Consequently, novel adsorption gels possessing high affinity or selectivity for platinum(IV) and palladium(II) were prepared by immobilizing various functional groups with high affinity for these metal ions onto polymer matrices of persimmon tannin extract by means of chemical modifications [6-9]. Scheme 2 shows the chemical structures of the chemically modified persimmon tannin extract gel prepared by the authors. In the present paper, we investigated the adsorption behaviors of the persimmon tannin extract gel immobilized with the functional groups of glycidyltrimethyl ammonium chloride, abbreviated as GTA-PT gel, which is a kind of polymeric quaternary ammonium chloride compounds, for platinum(IV) and palladium(II) from acidic chloride media.

Experimental

Preparation of the adsorbent

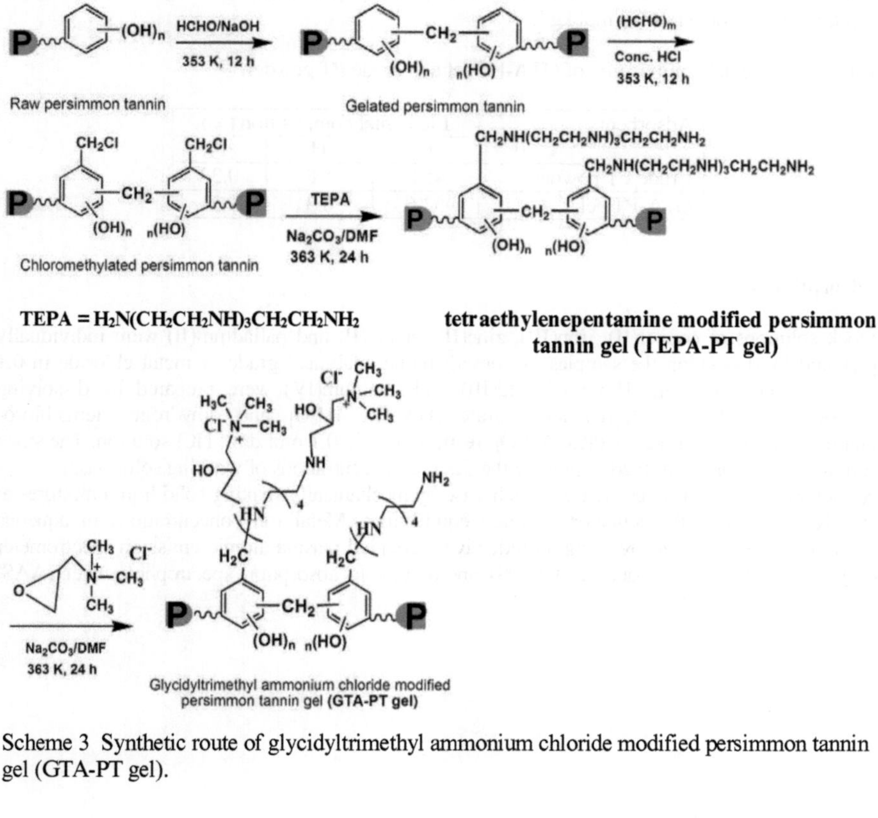

Scheme 3 Synthetic route of glycidyltrimethyl ammonium chloride modified persimmon tannin gel (GTA-PT gel).

Sample of dry persimmon tannin extract powder, the feed material, was kindly donated by Persimmon-Kaki Technology Development, Co. Ltd., Jincheng, China. This powder was commercially produced according to the processes as follows; 1) juicing of young astringent persimmon rich in persimmon tannin, 2) fermentation, and 3) conversion of the extract liquid into powder by spray drying. The GTA-PT gel was prepared from this powder according to the synthetic route as shown in Scheme 3 via two intermediate products, chloromethylated persimmon tannin and tetraethylenepentamine modified persimmon tannin (TEPA-PT).

The details of the preparation of these two intermediate products, chloromethylated persimmon tannin and tetraethylenepentamine modified persimmon tannin gel (TEPA-PT gel), are described elsewhere [7]. Four gram of TEPA-PT gel was suspended in 25 ml of dimethylformamide (DMF), to which 5 ml of glycidyltrimethyl ammonium chloride (DTA) was added to undergo the modification reaction in the presence of 2.0 g of sodium bicarbonate at 363 K for 24 h. The product was successively washed with distilled water followed by drying and grinding. Finally, it was sieved into uniform particle size up to 75 μm.
The elemental composition of thus prepared GTA-PT gel is shown in Table 1, compared with crude PT powder, the feed material. The density of nitrogen atom immobilized by the above-mentioned reaction was 6.65 mol kg^{-1}.

Table 1 Elemental composition of GTA-PT gel and crude PT powder.

Adsorbent	Elemental composition (%)		
	C	H	N
Crude PT powder	45.8	5.8	0.33
GTA-PT gel	53.6	6.3	9.31

Adsorption tests

Stock solutions of copper(II), iron(III), zinc(II), nickel(II) and palladium(II) were individually prepared by dissolving the samples of corresponding analytical grade of metal chloride in 0.1 mol dm^{-3} HCl solution. Those of gold(III) and platinum(IV) were prepared by dissolving hydrogen tetrachloroaurate(III) tetrahydrate [$HAuCl_4$ $4H_2O$] and dihydrogen tetrachloro-platinate(IV) hexahydrate [H_2PtCl_6 $6H_2O$], respectively, in 0.1 mol dm^{-3} HCl solution. The stock solutions were further diluted to prepare the desired concentrations of working solutions.
All adsorption tests were carried out batch-wise by mechanically shaking solid liquid mixtures at 303 K in an air bath incubator to attain equilibrium. Metal ion concentrations in aqueous solutions were measured by using an inductively coupled plasma atomic emission spectrometer (ICP-AES, Shimadzu model 8100 ICPS) and/or atomic absorption spectrophotometer (AAS, Shimadzu model 6800)

Results and Discussion

Effect of HCl concentration on the adsorption of metal ions on GTA-PT gel.

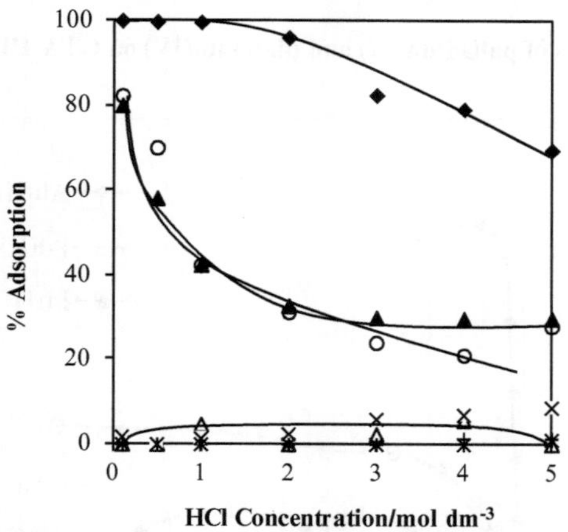

Figure 2 Plot of % adsorption of metal ions on GTA-PT gel vs. HCl concentration. Conditions: dry weight of the adsorbent = 10 mg, volume of the test solution = 10 ml, initial metal concentrations = 0.2 mmol dm^{-3}, shaking time = 48 h.

Figures 2 illustrates the effect of HCl concentration on % adsorption of each metal ion on GTA-PT gel. Here, the % adsorption is defined by the following equation.

$$\% \text{ Adsorption} = (C_i - C_e)/C_i \times 100$$

where C_i = initial metal concentration, C_e = equilibrium metal concentration.

As seen in this figure, nearly quantitative adsorption of gold(III) is achieved in the low concentration range of HCl and adsorption of palladium(II) and platinum(IV) is around 80 % at 0.1 mol dm^{-3} HCl while those of base metals such as copper(II) and iron(III) are negligible also in this case. That is, it is apparent that, by the immobilizing the functional groups of glycidyltrimethyl ammonium chloride onto polymer matrices of persimmon tannin, the adsorption of palladium(II) and platinum(IV) was much improved. The decrease in the adsorption of gold(III), palladium(II) and platinum(IV) with increasing HCl concentration is attributable to the anion exchange reaction of chloride ion paired with the functional groups of glycidyltrimethyl ammonium with chloride ion in the aqueous solution. Consequently, the adsorptive recovery of platinum(IV) and palladium(II) from acidic chloride is expected to be easily achieved using GTA-PT gel after the recovery of gold(III) using CPT gel or PP gel, for example.

Adsorption isotherms of palladium(II) and platinum(IV) on GTA-PT gel.

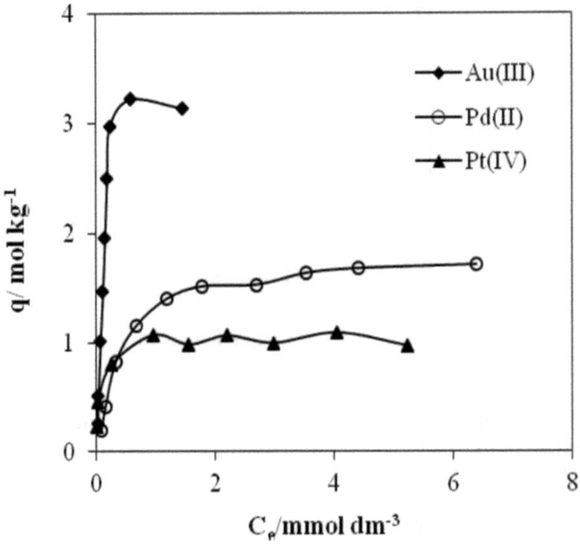

Figure 3 Adsorption isotherms of gold(III), palladium(II) and platinum(IV) on GTA-PT gel from 0.1 mol dm^{-3} HCl at 303 K.
Conditions: dry weight of the gel = 10 mg, volume of the test solutions = 10 ml, shaking time = 96 h.

Figure 3 shows the plot of the amount of adsorption of gold(III), palladium(II) and platinum(IV) on GTA-PT gel (q) vs. equilibrium metal concentration in the aqueous solution (Ce), adsorption isotherms, from 0.1 mol dm^{-3} HCl at 303 K. Here, the amount of adsorption is defined as follows.

$$q = (C_i - C_e) V / W$$

where C_i = initial metal concentration, C_e = equilibrium metal concentration, V = volume of metal solution, and W = dry weight of adsorbent

As seen from this figure, amount of adsorption rapidly increases with increasing metal concentration in the low concentration range while it tends to approach constant value in the high concentration range for all metal ions tested, suggesting that the adsorption takes place according to the Langmuir type adsorption. Consequently, the plots were rearranged according to the Langmuir equation as shown below.

$$\frac{Ce}{q} = \frac{1}{b\, q_{max}} + \frac{Ce}{q_{max}}$$

where b and q_{max} denote the Langmuir constant, or the adsorption equilibrium constant, and maximum adsorption capacity, respectively.

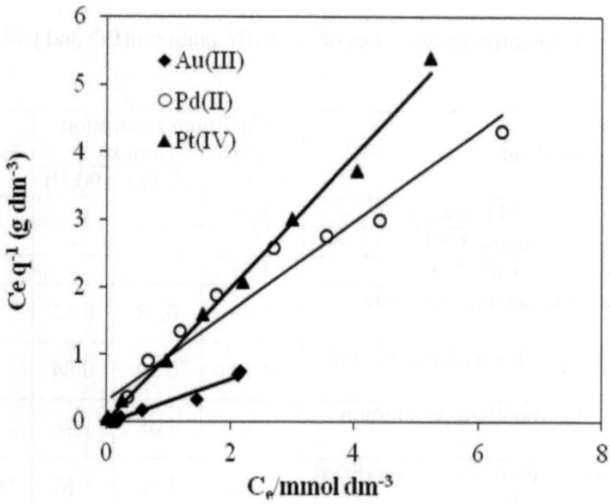

Figure 4 The Langmuir plots of the equilibrium adsorption amount for gold(III), palladium(II) and platinum(IV).

Figure 4 shows the Langmuir plot according to this equation, plot of Ce/q vs. Ce. As expected from this equation, all plots appear to lie on straight lines for each metal ion. From the slopes and intercepts with the ordinate of these straight lines, the maximum adsorption capacity, q_{max}, and the Langmuir constant, b, were evaluated according to the above-described equation. Table 2 shows thus evaluated values for each metal ion.

Table 2 Maximum adsorption capacities and adsorption equilibrium constants for GTA-PT gel.

Metal ion	Maximum adsorption capacity (q_{max}, mol kg^{-1})	Adsorption equilibrium constant (b, dm^3 mol^{-1})
Au(III)	3.35	14.0
Pd(II)	1.49	2.24
Pt(IV)	1.01	8.86

Table 3 shows the comparison of the maximum adsorption capacity for gold(III), palladium(II) and platinum(IV) of GTA-PT gel with persimmon tannin based adsorption gels evaluated in our previous works as well as those on other adsorbents reported in literatures. As shown in this table, it is noteworthy that the adsorption gels prepared from various biomass exhibits much higher adsorption capacities for gold(III) and platinum(IV) compared to a commercially available ion exchange resins, Duolite GT-73 and Lewatit TP 214, for example. The maximum adsorption capacities of GTA-TP gel for platinum(IV) and palladium(II) investigated in this work are higher than those of QATP gel, also containing functional groups of quaternary ammonium chloride, which may be attributable to the higher quantities of the immobilized quaternary amine groups.

Table 3 The maximum adsorption capacities of gold(III), platinum(IV) and palladium(II) on various adsorbents.

Adsorbent	Maximum adsorption capacity [mol kg^{-1}]			Reference
	Au(III)	Pt(IV)	Pd(II)	
Persimmon extract tannin (PT) powder	5.89	-	-	4
Cross-linked persimmon tannin (CPT)	7.7	-	-	4
Cross-linked persimmon waste(CPW)	4.95	-	-	5
Dimethylamine modified persimmon waste (DMA-PW)	5.63	0.28	0.42	10
Quaternary amine modified persimmon tannin (QAPT)	4.16	0.52	0.84	6
Tetraethylenepentamine modified persimmon tannin (TEPA-PT)	5.93	1.48	1.76	7
Glycidyltrimethyl ammonium chloride modified persimmon tannin (GTA-PT)	3.30	1.00	1.67	This work
Aminoguanidine modified persimmon tannin (AG-PT)	8.90	1.00	2.00	8
Bisthiourea modified persimmon tannin (BTU-PT)	5.22	0.70	1.72	9
Cross-linked microalgal residue	3.25	0.15	0.25	10
Quaternary amine modified microalgal residue	-	1.33	2.95	11
Cross-linked chitosan	-	1.6	2.1	12
Collagen fiber immobilized bayberry tannin	-	0.495	0.80	13
Lysine modified cross-linked chitosan	0.35	0.66	1.03	14
Glycine modified cross-linked chitosan	0.86	0.62	1.13	15
Ethylenediamine modified chitosan nanoparticle	-	0.87	1.30	16
Thiourea modified chitosan microsphere	-	0.16	1.06	17
Dimethylamine modified waste paper	4.60	0.90	2.10	18
Rice hulls	-	0.215	-	19
Lewatit TP 214	-	0.170	-	19
Duolite GT-73	0.58	0.26	-	20

Conclusion

Persimmon tannin based novel bio-adsorbent was prepared by immobilizing functional groups of glycidyltrimethyl ammonium chloride onto polymer matrices of persimmon tannin by chemical modification. This gel was found to exhibit high selectivity to platinum(IV) and palladium(II) over base metals such as copper(II) and iron(III). The adsorption mechanism of these metal ions from acidic chloride media on this gel was inferred to be interpreted in terms of the anion exchange reaction of chloride ion contained in this gel with anionic chloro-complexes of these metal ions. From the investigation of adsorption isotherms, it was found that the Langmuir type of adsorption takes on this gel and the maximum adsorption capacities evaluated for platinum(IV) and palladium(II) are higher than those of QAPT gel, the gel containing the functional group of trimethyl ammonium chloride, also functional groups of quaternary ammonium chloride, which may be attributable to the higher quantities of the immobilized quaternary amine groups.

References

1. Sakaguchi, T.; Nakajima, A., Recovery of uranium from seawater by immobilized tannin. *Sep.Sci.Technol.*, **1987**, 22, 1609-1623.
2. Nakano, Y.; Takeshita, K.; Tsutsumi, T., Adsorption mechanism of hexavalent chromium by redox within condensed-tannin gel, *Wat.Res.*, **2001**, 35, 496-500.
3. Matsuo, T.; Ito, S., The chemical structure of Kaki-tannin from immature fruit of the persimmon (*Diospyros kaki* L.), *Agric.Biol.Chem.*, **1978**, 42, 1637-1643.
4. Gurung, M.; Adhikari, B.B.; Kawakita, H.; Ohto, K.; Inoue, K.; Alam, S., Recovery of Au(III) by using low cost adsorbent prepared from persimmon tannin extract, *Chem.Eng.J.*, **2011**, 174, 556-563.
5. Parajuli, D.; Kawakita, H.; Inoue, K.; Ohto, K.; Kajiyama, K., Persimmon peel gel for the selective recovery of gold, *Hydrometallurgy*, **2007**, 87, 133-139.
6. Gurung, M.; Adhikari, B.B.; Khunathai, K.; Kawakita, H.; Ohto, K.; Harada, H.; Inoue, K., Quaternary amine modified persimmon tannin gel: An efficient adsorbent for the recovery of precious metals from hydrochloric acid media, *Sep.Sci.Technol.*, **2011**, 46, 2250-2259.
7. Gurung, M.; Adhikari, B.B.; Alam, S.; Kawakita, H.; Ohto, K.; Inoue, K., Persimmon tannin-based new sorption material for resource recycling and recovery of precious metals, *Chem.Eng.J.*, **2013**, 228, 405-414.
8. Gurung, M.; Adhikari, B.B.; Morisada, S.; Kawakita, H.; Ohto, K.; Inoue, K.; Alam, S., N-amonoguanidine modified persimmon tannin: A new sustainable material for selective adsorption, preconcentration and recovery of precious metals from acidic chloride solution, *Biores.Technol.*, **2013**, 129, 108-117.
9. Gurung, M.; Adhikari, B.B.; Kawakita, H.; Ohto, K.; Inoue, K.; Alam, S., Selective recovery of precious metals from acidic leach liquor of circuit boards of spent mobile phones using chemically modified persimmon tannin gel, *Ind.Eng.Chem.Res.*, **2012**, 51, 11901-11913.

10. Khunathai, K.; Xiong, Y.; Biswas, B.K.; Adhikari, B.B.; Kawakita, H.; Ohto, K.; Inoue, K.; Kato, H.; Kurata, M.; Atsumi, K., Selective recovery of gold by simultaneous adsorption-reduction using microalgal residues generated from biofuel conversion processes, *J.Chem.Technol.Biotechnol.*, **2012**, 87, 393-401.
11. Khunathai, K.; Inoue, K.; Ohto, K.; Kawakita, H.; Kurata, M.; Atsumi, K.; Fukuda, H.; Alam, S., Adsorpotive recovery of palladium(II) and platinum(IV) on the chemically modified-microalgal residue, *Solv.Extr.Ion Exch.*, **2013**, 31, 320-334.
12. Inoue, K.; Baba Y.; Yoshizuka, K., Adsorption of metal ions on chitosan and crosslinked copper(II)-complexed chitosan. *Bull.Chem.Soc.Japan*, **1993**, 66, 2915-2921.
13. Wang, R.; Liao, X.; Shi, B., Adsorption behaviors of Pt(IV) and Pd(II) on collagen fiber immobilized bayberry tannin, *Ind.Eng.Chem.Res.*, **2005**, 44, 4221-4226.
14. Fujiwara, K.; Ramesh, A.; Maki, T.; Hasegawa, H.; Ueda, K., Adsorption of platinum(IV), palladium(II) and gold(III) from aqueous solutions on L-lysine modified crosslinked chitosan resin, *J.Hazard.Mater.*, **2007**, 146, 39-50.
15. Ramesh, A.; Hasegawa, H.; Sugimoto, W.; Maki, T.; Ueda, K., Adsorption of gold(III), platinum(IV) and palladium(II) onto glycine modified crosslinked chitosan resin, *Biores.Technol.*, **2008**, 99, 3801-3809.
16. Zhou, L.; Xu, J.; Liang, X.; Liu, Z., Adsorption of platinum(IV) and palladium(II) from aqueous solutions by magnetic crosslinking chitosan nanoparticles modified with ethylenediamine, *J.Hazard.Mater.*, **2010**, 182, 518-524.
17. Zhou, L.; Liu, J.; Liu, Z., Adsorption of platinum(IV) and palladium(II) from aqueous solution by thiourea-modified chitosan microspheres, *J.Hazard.Mater.*, **2009**, 172, 439-446.
18. Adhikari, C.R.; Parajuli, D.; Kawakita, H.; Inoue, K.; Ohto, K.; Harada, H., Dimethylamine-modified waste paper for the recovery of precious metals, *Env.Sci.Technol.*, **2008**, 42, 5486-5491.
19. Morcali, M.H.; Zeytuncu, B.; Yucel, O., Comparison of the adsorption by rice hulls and Lewatit TP 214 of platinum from chloroplatinic solution, *J.Serb.Chem.Soc.*, **2013**, 78, 811-826.
20. Iglesias, M.; Antico, E.; Salvado, V., Recovery of palladium(II) and gold(III) from diluted liquors using the resin Duolite GT-73, *Anal.Chim.Acta*, **1999**, 381,61-67.

Rare Metal Technology 2016
Edited by: Shafiq Alam, Hojong Kim, Neale R. Neelameggham, Takanari Ouchi, and Harald Oosterhof
TMS (The Minerals, Metals & Materials Society), 2016

INVESTIGATION OF IRON REMOVAL FROM REDUCED UPGRADED TITANIA SLAG USING MILD ACIDS

Jaehun Cho, Syamantak Roy, Amarchand Sathyapalan, Michael L. Free, Zhigang Zak Fang

Department of Metallurgical Engineering, University of Utah, Salt Lake City, Utah 84112

Keywords: upgraded titania slag, iron removal, hydrochloric acid, titanium hydride

Abstract

Leaching characteristics of impurity removal from the Upgraded Titania Slag (UGS) using hydrochloric and boric acids has been investigated. Specifically, this study has focused on removal of iron oxides from the UGS that has been reduced by hydrogen gas and magnesium at 750°C for 5 hours. Reduced UGS is mainly composed of titanium hydride and impurities like iron oxides, silicon oxide, aluminum oxide, and magnesium oxide. The concentration of hydrochloric and boric acids and the leaching temperature have been chosen as the prime variables to identify the optimum conditions and leaching characteristics for removal of iron oxides. It was observed that under the constraint that 1 wt% of titanium hydride loss is the maximum amount tolerable, 0.1M HCl with 1M boric acid and 90°C are the most effective parameters for iron removal from the reduced UGS. Also, the overall rate of reaction seems to follow interfacial reaction controlled kinetics. The calculated activation energy of the reaction is 73.90kJ/mol.

1. Introduction

Titanium is an important chemical material widely used in aviation, aerospace, biomedical, marine, and nuclear waste storage application due to its corrosion resistance, high strength and light weight [1]. Although it is predominantly found as rutile (TiO_2) and ilmenite ($FeTiO_3$), which are abundant in the earth's crust, the production cost is relatively high compared to competing metals such as steel and aluminum. The main method for extraction of titanium is the Kroll process in which metallic titanium is obtained by the reduction of titanium tetrachloride ($TiCl_4$), which is produced from the chlorination of titanium dioxide or synthetic rutile, in the presence of magnesium [2]. Even though this process allows industry to minimize the amount of impurities in the metallic titanium, it causes high production costs and environmental issues such as generation of greenhouse gases and corrosive intermediate product like titanium tetrachloride [3-5].

To mitigate the challenges of the Kroll process several new alternative techniques have been suggested. Dr. Fang et al. (2013) has investigated titanium production from titanium hydride, which can be obtained by reduction of titania slag or synthetic rutile under a controlled environment in the presence of magnesium and salts [6]. In this new route, production costs and environmental problems can be significantly reduced compared to the Kroll process, but the impurities such as iron in the final product can be challenging to remove [6].

This study is about the hydrometallurgical leaching characteristics associated with the removal of iron oxides from upgraded titania slag (UGS) which has been reduced using hydrogen gas and magnesium at 750°C for 5 hours [6]. Reduced UGS is predominantly composed of titanium hydride and impurities like iron oxides, silicon oxide, aluminum oxide and magnesium oxide. Mild concentrations of hydrochloric and boric acids were chosen as lixiviants and their concentration and leaching temperature were considered as prime variables to identify the optimum condition and leaching characteristics for iron oxides removal.

2. Experimental Procedures

2.1. Preparation of the Reduced UGS

UGS is upgraded titania slag composed of about 96% titanium dioxide and 4% impurities. Mixture of UGS (less than 40um), magnesium powder and salt like magnesium chloride and potassium chloride with a ratio of 1:1:1 by weight are put in the furnace under hydrogen atmosphere and reduced for 5 hours at the temperature of 750°C. After that, the product resulting from the reduction is washed using 4.3M acetic acid with a solid liquid ratio of 1g to 40ml for 2 hours at the temperature of 70°C to remove magnesium compounds and salts from the reduced UGS, the composition of which after acetic acid leaching is shown in the Table 1.

Elements	MgO	TiH_2	Fe_2O_3	Al_2O_3	SiO_2
wt %	0.36	96.20	1.07	0.35	2.02

Table 1. Composition of reduced UGS after acetic acid washing by ICP-OES

Figure 1 presents the mineralogical phases of the acetic acid washed, reduced UGS analyzed by XRD from which pattern shows that the crystalline phase of products predominantly consist of titanium hydride and ferric oxide.

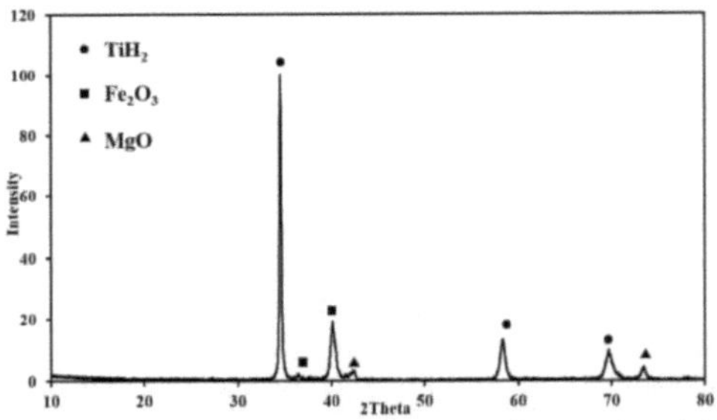

Figure 1. XRD pattern of the reduced UGS after acetic acid washing

After reduction, the product was mixed by a mixing machine for 30 minutes. The particle size of the reduced UGS was in the range of 0.05 – 0.2um.

2.2. Experimental Procedure

200ml of solution which contains mild hydrochloric and boric acids was prepared in the flask made of semi-clear polymethylpentene with 2g of the reduced UGS after acetic acid washing. A chemical and temperature resistant plug with a hole that allows for a thermometer insertion is applied on the top of the flask to maintain constant volume. The solution is mixed by a magnetic stirring bar at a rotational speed of 1,000RPM and heated to the predetermined temperature. The leaching experiment is run for 4hours and samples are collected every hour for the study of reaction kinetics. The samples were analyzed by Inductively Coupled Plasma-Optical Emission Spectroscopy (ICP-OES).

One of the difficulties in the leaching process of the reduced UGS is that titanium hydride is easily oxidized to titanium dioxide under a severe environment. Especially, titanium hydride tends to be converted to titanium chloride under the strong concentration of hydrochloric acid. Thus, before conducting the regular experiments, the acceptable amount of titanium hydride loss during the leaching process needs to be decided. Our research group determined arbitrarily that 1 weight percent of titanium loss is the maximum amount tolerable. Test results for titanium loss as a function of HCl concentration is presented in Figure 2.

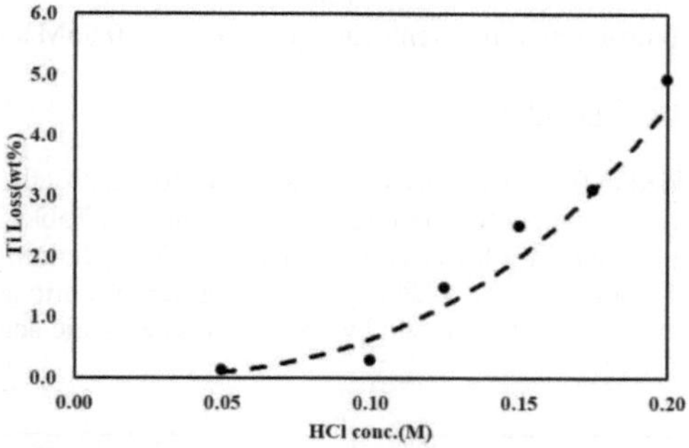

Figure 2. Titanium loss versus HCl concentration

Based on the results in Figure2, the maximum concentration of HCl that meets the requirement of less than 1% titanium loss is approximately 0.1M. The boric acid concentration was selected as 1M.

3. Results and Discussion

3.1. Effect of HCl Concentration

The influence of HCl concentration on the removal of iron oxide was studied with 0.05M and 0.1M HCl under the following leaching parameters in Table 2.

Temp.	Time	Stirring speed	Solid-liquid ratio	Particle size
90°C	4hrs	1,000RPM	1g : 100ml	0.05 – 0.2um

Table 2. Leaching parameters

Figure 3 presents the effect of HCl concentration on iron removal. 0.1M HCl shows 55% of iron removal after 4 hours. Although the concentration of HCl significantly affects iron removal, the maximum HCl concentration is restricted to reduce the titanium loss. This leads to the introduction of new additive like boric acid, which is known as an effective acid for removal of hydrated iron oxide.

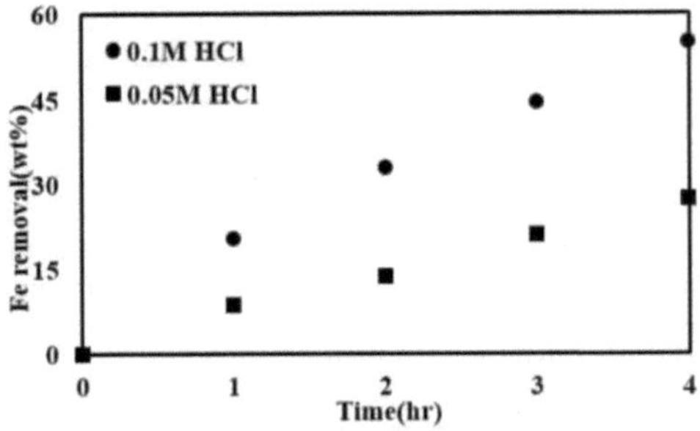

Figure 3. Comparison of iron removal versus between 0.05M and 0.1M HCl

3.2. Effect of Boric Acid Concentration

1M boric acid was added to 0.05M and 0.1M HCl solutions to identify the effect of boric acid on iron removal. Leaching parameters for these tests are presented in Table 2. Figure 4a on the left shows the effect of boric acid in a lower concentration of HCl solution. The iron removal rate significantly increased from 27.3% to 43.8% with the increase of boric acid. However, in 0.1M HCl solution, the iron removal rate increased by only 6.4% when boric acid is added according to Figure 4b on the right.

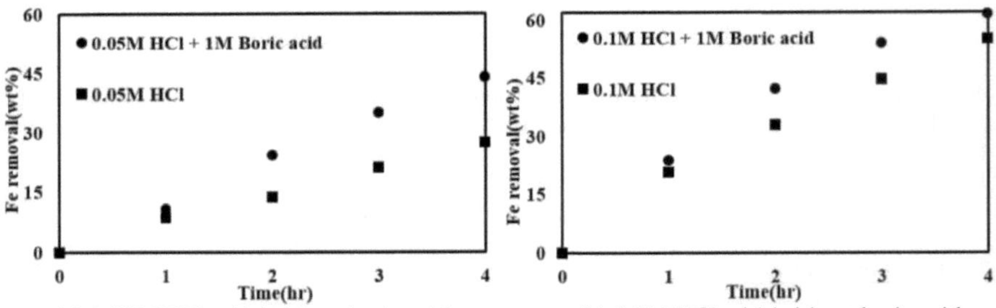

(a) 0.05M HCl with/without boric acid (b) 0.1M HCl with/without boric acid

Figure 4. Comparison of iron removal versus time with/without boric acid

3.3. Effect of Temperature

The effect of leaching temperature on iron removal was studied in the temperature range of 70 - 90°C, with 0.1M HCl solution under leaching parameters in Table 2. Figure 5 shows that iron extraction is strongly related to leaching temperature. Iron removal rate increases rapidly with the increase of temperature.

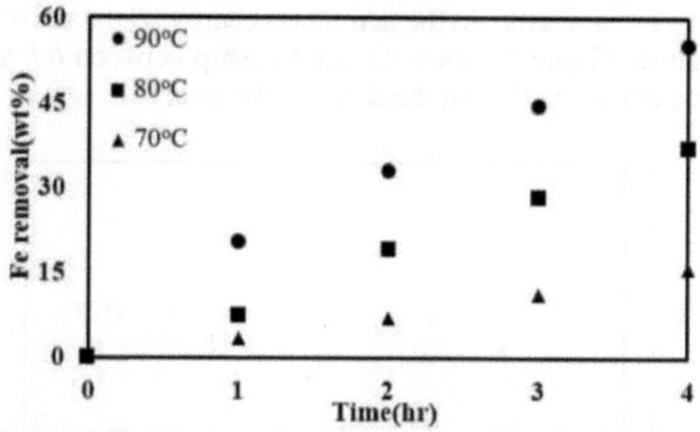

Figure 5. Iron removal versus time at three temperatures

The rate-controlling process and kinetic parameters were investigated on the basis of a reaction controlled kinetic model. If the overall rate of reaction is controlled by the interfacial reaction, X can be described as follows [7-9]:

$$kt^* = 1 - (1 - X)^{1/3} \tag{1}$$

where k is reaction rate constant and t^* is time which incorporates such system parameters as equilibrium constant, shape factor of particle, bulk fluid concentration, and stoichiometry. The experimental data was applied to the equation for interfacial reaction controlled kinetics as illustrated by plotting $1 - (1 - X)^{1/3}$ versus time as shown in Figure 6.

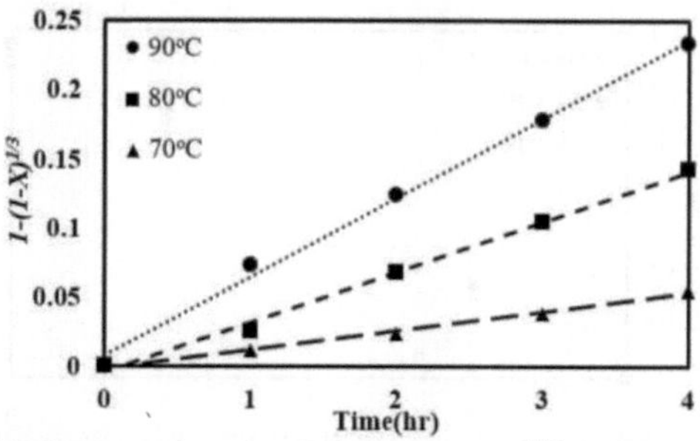

Figure 6. Relationship between the values of $\mathbf{1 - (1 - X)^{1/3}}$ and time at three temperatures

The value of $1 - (1 - X)^{1/3}$ is linearly related to leaching time, which suggests that the overall rate of reaction may be controlled by the interfacial reaction in the leaching process. Reaction rate constants at three different temperature in Equation 1 are obtained from the slope of linear lines in Figure 6. The activation energy can also determine the rate-controlling step. The Arrhenius equation, when rearranged for determining the activation energy is described in Equation 2.

$$lnk = lnA - \frac{E_a}{RT} \tag{2}$$

where A is the pro-exponential factor, E_a the activation energy, R the universial gas constant, and T the absolute temperature. Figure 7 shows the relationship between lnk and $10^3/T$ in which the activation energy can be determined from the slope of the line.

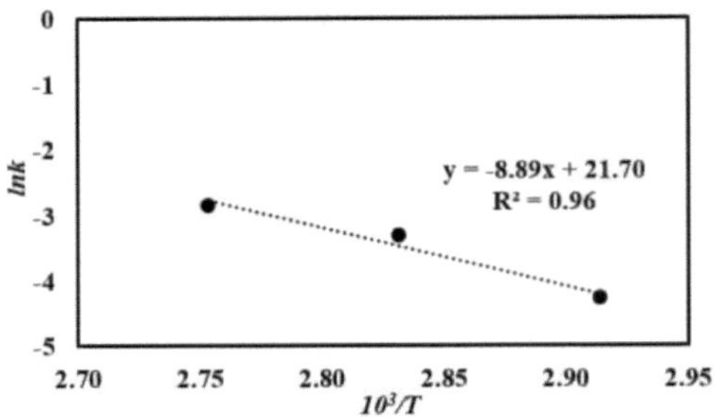

Figure 7. Arrhenius Plot

The activation energy is 73.90kJ/mol which suggests that interfacial reaction controlled kinetics can describe the observed data.

3.4. Effect of Mass Transport

Influence of mass transport on the reaction rate of leaching with 0.1M HCl was studied by varying the stirring speed in the range of 100 – 1000 rpm under the leaching parameters in Table 2.

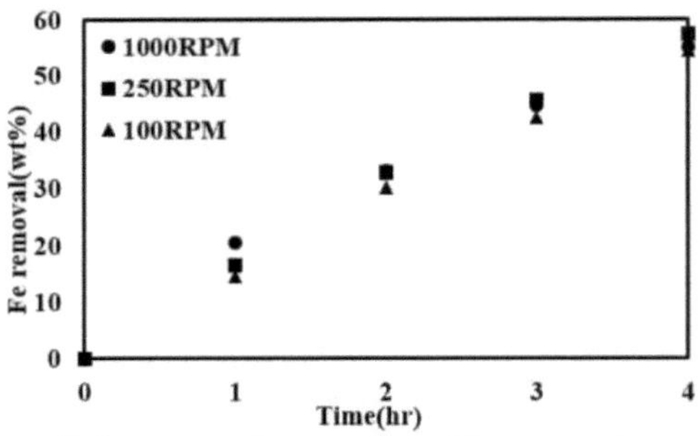

Figure 8. Iron removal versus time at three mass transports

The data in Figure 8 show that the reaction rate did not change significantly with increasing stirring speed, which implies that the overall rate of reaction is not controlled by mass transport.

4. Conclusions

The leaching characteristics of iron removal from the reduced upgraded titania slag with mild hydrochloric and boric acids was studied.

(1) Under the constraint that 1 wt % of titanium hydride loss is the maximum amount tolerable, 0.1M HCl is the most effective lixiviant for iron removal from the reduced UGS.
(2) 1M boric acid helps significantly remove iron from the reduced UGS with 0.05M HCl. However, with 0.1M HCl, 1M boric acid helps increase iron removal by only 6.4% compared to the case where there is no boric acid in the solution.
(3) Iron removal rate increases rapidly with the increase of temperature.
(4) The overall rate of reaction seems to follow interfacial reaction controlled kinetics. The activation energy of the reaction is 73.90kJ/mol.
(5) Change in mass transport does not affect the reaction rate significantly.

References

[1] S. Amarchand, T.R. Rama Mohan, and P. Ramakrishnan, "A Novel Chemical Solution Technique for the Preparation of Nano Size Titanium Powders from Titanium Dioxide", Advanced Powder Technol., 11 (4) (2000), 415-422.
[2] W. J. Kroll, "The Production of Ductile Titanium", Transactions of the Electrochemical Society, 78 (1940), 35–47.
[3] N. El-Hazek, T.A. Lasheen, R. El-Sheikh, and S.A. Zaki, "Hydrometallurgical Criteria for TiO_2 leaching from Rosetta ilmenite by hydrochloric acid", Hydrometallurgy, 87 (2007), 45-50.
[4] S. Middlemas, Z.Z. Fang, and P. Fan, "A New Method for Production of Titanium Dioxide Pigment", Hydrometallurgy, 131-132 (2013), 107-113.
[5] A. Mehdilo and M. Irannajad, "Iron Removing from Titanium Slag for Synthetic Rutile Production", Physicochem. Probl. Miner. Process., 48 (2) (2012), 425-439.
[6] Z.Z. Fang, S. Middlemas, P. Fan, and J. Guo, "A Novel Chemical Pathway for Producing Low Cost Ti by Direct Reduction of Ti-slag", Journal of American Chemical Society, 135 (49) (2013), 18248–18251.
[7] M.L. Free, *Hydrometallurgy: Fundamentals and Applications*, (New Jersey: John Wiley & Sons, lnc., Hoboken, 2013).
[8] J. Szekely, J.W. Evans, and H.Y. Sohn, *Gas-Solid Reaction*, (New York, NY: Academic Press, 1976).
[9] L. Sui and Y. Zhai, "Reaction Kinetics of Roasting High-Titanium Slag with Concentrated Sulfuric Acid", Trans. Nonferrous Met. Soc. China, 24 (2014), 848-853.

Rare Metal Technology 2016
Edited by: Shafiq Alam, Hojong Kim, Neale R. Neelameggham, Takanari Ouchi, and Harald Oosterhof
TMS (The Minerals, Metals & Materials Society), 2016

Production of Tungsten by Pulse Current Reduction of $CaWO_4$

Furkan Özdemir[1], Metehan Erdoğan[2], Mustafa Elmadağlı[3], İshak Karakaya[1]

[1]Middle East Technical University
Department of Metallurgical and Materials Engineering;
Üniversiteler Mah. Dumlupınar Blv. No:1, 06800, Çankaya, Ankara/TURKEY

[2]Yıldırım Beyazıt University
Department of Materials Engineering;
Çankırı Caddesi Çiçek Sokak No:3, 06050 Ulus Altındağ/Ankara/TURKEY

[3]Roketsan
Kemalpaşa Mahallesi Şehit Yüzbaşı Adem Kutlu Sokak No:21
06780 Elmadağ, Ankara/TURKEY

Keywords: Electrodeoxidation, Tungsten, Scheelite, $CaWO_4$, Pulse Current

Abstract

Tungsten is mainly produced by H_2 reduction of WO_3 which makes its production difficult and costly. A promising method was recently reported for direct electrochemical production of tungsten from $CaWO_4$ (scheelite) which is estimated as the seventy percent of the all tungsten reserves of the world. This paper investigates the application of pulse current in the electrochemical reduction of $CaWO_4$ pellets attached to a current collector. The electrolyte was composed of calcium chloride and sodium chloride salts at eutectic composition (48 pct mol NaCl at 873 K). The reduced samples were cleaned with dilute hydrochloric acid solutions and characterized by X-ray diffraction and scanning electron microscopy. Examination of the results revealed that metallic tungsten was obtained by pulse current reduction.

Introduction

Tungsten is the highest melting point (3422±15° C) metal. Tungsten also has high density, very high thermal creep resistance, good high-temperature mechanical properties, very low thermal expansion coefficient, very high moduli of compression and elasticity and good conductivity for heat and electricity [1]. Having these significant properties makes tungsten important in many technological applications. Tungsten is mainly used in the production of cemented carbides which have high hardness and enables shaping of metals, alloys, wood, composites, plastics and ceramics, as well as for the mining and construction equipment. Other uses of tungsten are generally in chemical applications and products such as catalysts, pigments, paints and lubricants. In addition, superalloys and tungsten heavy metal alloys contain tungsten because of its outstanding properties [2].

There are more than thirty known tungsten containing minerals. However, only two of them have considerable industrial use, namely scheelite ($CaWO_4$) and wolframite ($(Fe, Mn)WO_4$). Scheelite is the most abundant mineral of tungsten and is present in approximately two thirds of known tungsten deposits [3]. The production of tungsten from the minerals is carried out by hydrogen reduction of tungsten oxides which are obtained from abovementioned minerals in the temperature range 600 to 1000°C. The conversion of wolframite to tungsten trioxide via alkaline dissolution processes is easier than scheelite. Despite the larger deposits of scheelite, the difficulties encountered in the alkaline dissolution process to transform scheelite to tungsten trioxide severely restrict the widespread use of scheelite in tungsten production [4]. Moreover; hydrogen reduction has many disadvantages. Besides the danger of explosion, it requires continuous external heat supply to attain and keep the high temperatures required for endothermic reduction reaction [5].

An alternative method was recently reported for direct electrochemical production of tungsten from $CaWO_4$ [6-8]. This method claims the electrochemical reduction of calcium tungstate to tungsten by using a $NaCl–CaCl_2$ eutectic electrolyte at 873 K. The resulting product was the mixture of tungsten metal and calcium compounds [8]. Calcium compounds were removed by dissolution in dilute hydrochloric acid. The studies [4,9] revealed that electrochemical method has the potential to decrease energy consumption in tungsten production. Furthermore, pulse potential applications were reported to have further flexibility to influence electrode processes [10]. Since it is known that pulse current plating improves mass transport, electrode kinetics and the nucleation of growth centers [11], its application to electrochemical reduction of $CaWO_4$ was studied in this investigation.

Experimental

2,5 grams of a commercially available $CaWO_4$ powder (Aldrich 24,866) was dry pressed under 2 tons of load to form a pellet of about 1,5 cm diameter and 0.3 cm height. This $CaWO_4$ pellet was placed into a stainless steel spoon, forming the cathode was welded to the end of a 3 mm diameter stainless steel wire. A 13 mm diameter and about 18 cm long graphite rod (Alfa Aesar A10134) assembled to the end of a treaded stainless steel wire was used as the anode. A schematic drawing of the experimental set-up showing the details of the cell is given in Figure 1. The cell was positioned into a quartz vessel of 60 cm height and 5 cm inner diameter. Finally, the cell was covered by a castermid lid that was specially designed to insulate the top of the cell.

The electrolyte was composed of calcium chloride and sodium chloride salts at eutectic composition (48 %mol NaCl at 873 K). Both salts were dried to obtain anhydrous oxide free electrolyte. $CaCl_2$ was first heated to 110°C and held for three hours. Then the temperature was ramped to 500°C and held for 24 hours. All the drying process was conducted under vacuum to remove chemically bounded moisture and minimize oxide formation. On the other hand NaCl was heated to 110°C and held for 24 hours.

The quartz vessel holding the electrolytic cell was positioned into a vertical furnace and heated to the target temperature of 600°C at which the electrolyte was in molten state [12]. The electrodes were immersed into the electrolyte after melting was completed. The electrolysis experiments were conducted by applying pulse currents via a Plating Electronic pe86CB power supply. The average voltage data during pulse current application were recorded by a self-developed computer software and an Agilent U1273AX digital multimeter. Argon gas was supplied during heating and electrolysis experiments.

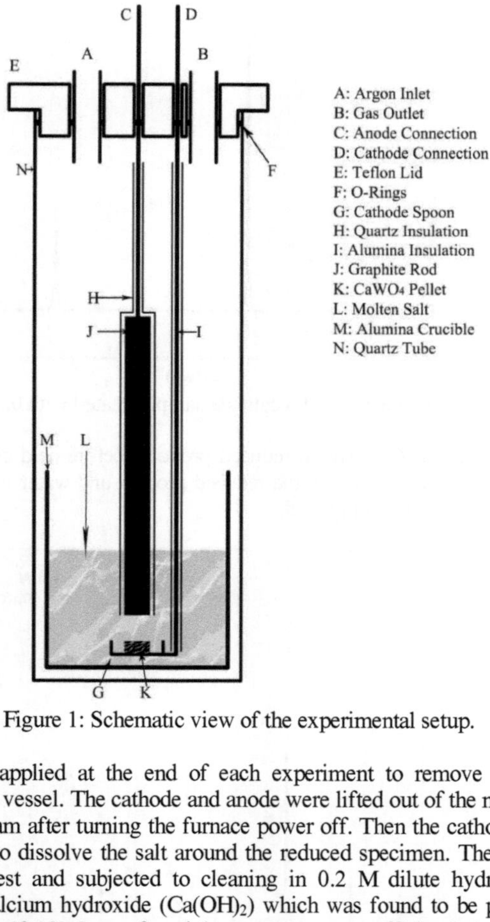

Figure 1: Schematic view of the experimental setup.

Argon flushing was applied at the end of each experiment to remove gaseous electrolysis products from the cell vessel. The cathode and anode were lifted out of the molten salt and left to cool in the argon stream after turning the furnace power off. Then the cathode was treated in an ultrasonic water bed to dissolve the salt around the reduced specimen. The solid particles were separated from the rest and subjected to cleaning in 0.2 M dilute hydrochloric acid (HCl) solution to remove calcium hydroxide ($Ca(OH)_2$) which was found to be present in significant amounts at the cathode [8,13]. It was found that, tungsten was sufficiently inert in such a solution while $Ca(OH)_2$ could easily and completely be removed. After the acid treatment, the sample was again filtered and dried. Reduced samples were investigated via X-ray diffraction analysis and scanning electron microscopy to analyze chemical compositions.

Results and Discussion

Figure 2 shows the result of an X-Ray diffraction of powder obtained from one of the experiments after HCl cleaning. Similar diffractions obtained from other pulse current applications indicated almost complete reduction of $CaWO_4$ pellets.

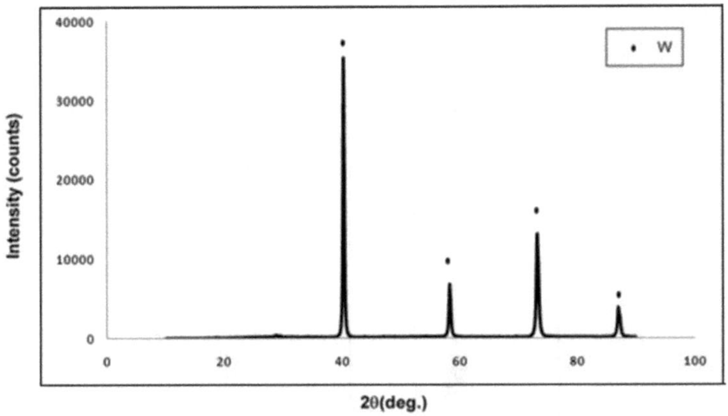

Figure 2: X-ray diffraction result of a cathode sample treated with 0.2 M HCl solution.

The presence of byproduct; $Ca(OH)_2$ in reduced products before acid cleaning can be seen in Figure 3. The reaction between CaO in the reduced product and water used to dissolve the salt resulted in the formation of this compound.

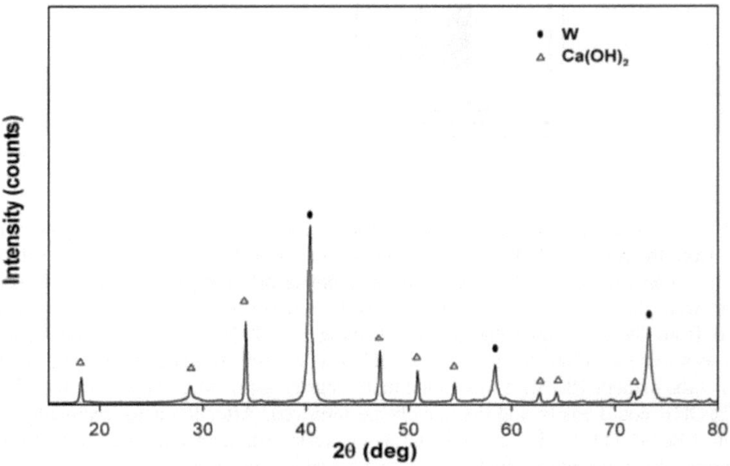

Figure 3: X-ray diffraction result of a cathode sample before acid cleaning.

All the characteristic peaks of tungsten were detected in the cleaned sample. SEM and EDS enable further analyses and characterizations of the reduced samples. SEM images of the end-products are presented in Figure 4 and EDS analysis of a selected sample is given in Table 1. The final particle size of the tungsten powders was observed to be less than 100 nm. EDS results indicated that the reduced samples were essentially composed of tungsten metal at a high percentage. Extended reduction of the pellets at lower currents showed complete reduction. Table 1 does not show oxygen concentration of the sample, because it was intentionally excluded from the assemblage to follow the correct proportions of the metallic constituents.

Figure 4.SEM photographs of the (a) nonsintered $CaWO_4$ pellet and (b) electrochemically reduced sample showing tungsten powder.

Table 1. EDS result of electrochemically reduced sample

Element	Wt %	At %	K-Ratio	Z	A	F
W M	98,50	93,46	0,9823	0,9942	1,0031	1,0000
Ca K	1,50	6,54	0,0106	1,3100	0,5375	1,0037
Total	100,00	100,00				

Experiments were conducted according to conditions given in the Table 2 and voltage values were recorded during the experiments as a function of time. The voltage vs. time graphs given in Figure 4 recorded for experiments demonstrated the typical shape of curve for constant current experiments [13]. In the experiments, average currents were set to constant values of 0.2 and 0.3 amperes. The corresponding peak currents and t_{on} and t_{off} durations are also given in Table 2. Applied potentials were not allowed to exceed 3.2 V which is the decomposition voltage of $CaCl_2$. Average currents were calculated according to (1)

$$I = \frac{I_{peak}}{t_{on} + t_{off}} \qquad (1)$$

Table2: Experimental Conditions and Results of Pulse Current Applications

Average Current, I	I_{peak}(A)	t_{on} (ms)	t_{off} (ms)	Energy Consumption (kWh/kg W)	Current Efficiency (%)
0.2	2	1	9	2.456	83.67
0.2	2.2	1	10	2.707	81.82
0.2	4	1	19	...	...
0.2	1	1	4	2.579	80.98
0.3	3	1	9	2.669	76.77
0.3	6	1	19	2.934	74.72

The accumulative (total) charge, Q, passing through the cell can be calculated from the below formula:

$$Q=\int_0^t I\,dt = \frac{100\,W z F}{A X} \qquad (2)$$

where I is the current in amperes, t is the time in seconds, F is the Faraday constant, z is the valence, A is the molecular weight of $CaWO_4$ and X is the percent current efficiency. About 5025 coulombs of charge should be passed to reduce 2.5 grams of $CaWO_4$ completely, when 100% current efficiency is assumed.

The total charge passed during each experiment, Q_a, was calculated from the average currents and the durations of each experiment. If "m" grams of tungsten are collected during this period, percent current efficiency, %CE, can then be calculated from the ratio of Q_c, the theoretical charge required for "m" grams of tungsten, to Q_a.

$$\%CE = \frac{Q_c}{Q_a} * 100 \qquad (3)$$

%CE determined from equation (3) for each run are listed in Table 2. The applied potential versus time data given in Fig. 5, and the average currents can be used to calculate energy consumption, ε, for each experiment according to:

$$\varepsilon = \int_0^t I * V\,dt \qquad (4)$$

The amount of cleaned samples collected after each experiment and their purities (95 to 99 wt.% W) together with the energy consumption, ε, were used to determine, E (per kg W), energy required per kilogram of tungsten production;

$$E(per\ kg\ W) = \frac{1000}{m} x\,\varepsilon \qquad (5)$$

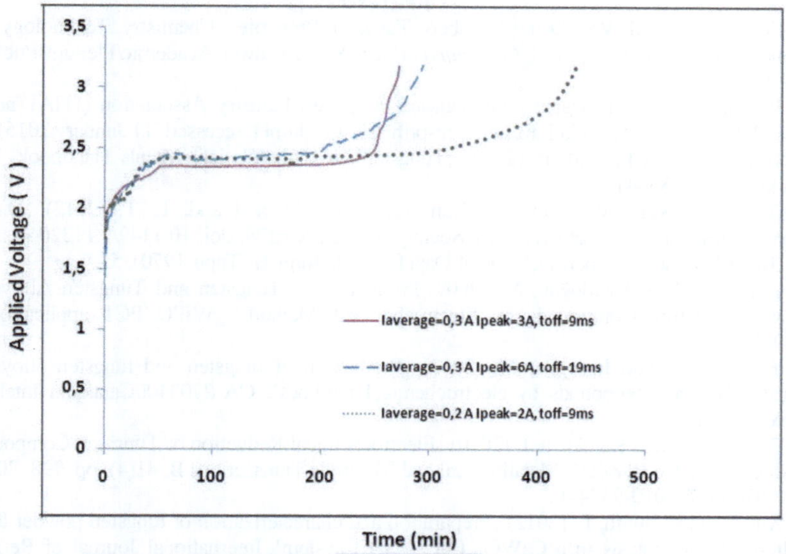

Figure 5.Applied potential vs. time graphs of selected experiments.

E(per kg W) determined from equation (5) for each run are listed in Table 2. The results indicated that energy consumption values of the process were between 2.45 and 2.93 kWh per kg tungsten metal powder. There was one sample (1 ms t_{on} and 19 ms t_{off} durations and 0.2 average applied current) for which results are not given in Table 2. The reduction reaction was very slow compared to other samples and calculated current efficiency and energy consumption had large errors due to very long reduction periods. All the other results indicated that pulse current studies were more energy efficient than constant voltage studies [4,9].

Conclusions

The effects of pulse current application on the electrochemical reduction of $CaWO_4$ pellets in eutectic molten salt mixture ($CaCl_2$-NaCl) at 600°C were investigated. It has been shown that reduction of the $CaWO_4$ pellets were successfully completed by pulse currents. Analyses of data obtained from pulse current applications revealed lower energy consumptions and higher current efficiencies than the constant voltage studies.

Acknowledgements

The authors acknowledge the financial support provided by the Ministry of Science, Industry and Technology of Turkey and Roketsan Incorporation. The contents of this paper are the sole responsibility of the authors and the ideas and opinions presented do not necessarily reflect those of Ministry of Science, Industry and Technology of Turkey and Roketsan Incorporation.

References

[1] Erik Lassner and Wolf-Dieter Schubert, Tungsten Properties, Chemistry, Technology of the Element, Alloys, and Chemical *Compounds* (New York, Kluwer Academic/Plenum Publishers, 1999), 7-42.

[2] Primary Uses of Tungsten | International Tungsten Industry Association (ITIA) (no date) Available at: http://www.itia.info/tungsten-primary-uses.html (Accessed: 11 January 2015).

[3] Brown, T. and Pitfield, P. (2013) „Tungsten', in Gunn/Critical Metals Handbook. Wiley-Blackwell, pp. 385–413.

[4] Tang, D., Xiao, W., Yin, H., Tian, L., Wang, D. and and, L. T. (2012) „Dingding Tang', Journal of The Electrochemical Society, 159(6), p. E139. doi: 10.1149/2.113206jes.

[5] TerkelRosenqvist, Thermochemical Data for Metallurgists (Tapir 1970), 55.

[6] Karakaya, İ. and Erdoğan, M. (2009)„Production of Tungsten and Tungsten Alloys from Tungsten Bearing Compounds by Electrochemical Methods', WIPO PCT application WO 2009/054819A1.

[7] Karakaya, İ. and Erdoğan, M. (2013) „Production of tungsten and tungsten alloys from tungsten bearing compounds by electrochemical methods', CA 2703400Canadian Intellectual Property Office.

[8] Erdoğan, M. and Karakaya, İ. (2010) „Electrochemical Reduction of Tungsten Compounds to Produce Tungsten Powder', Metallurgical and Materials Transactions B, 41(4), pp. 798–804. doi: 10.1007/s11663-010-9374-4..

[9] Xu, W. and Chunfa, L. (2012) „Preparation and characterization of tungsten powder through molten salt electrolysis in a $CaWO_4$–$CaCl_2$–NaCl system', International Journal of Refractory Metals and Hard Materials, 31, pp. 205–209. doi: 10.1016/j.ijrmhm.2011.11.004.

[10] Göksu, O.G.,Karakaya, İ. and Erdoğan, M. (2010)„Constant and Pulse Voltage Applications in CaWO4 Reduction', EPD Congress, 387-394 ed. Edgar E. Vidal,"TMS", Washington.

[11] Gomes, A. and da Silva Pereira, M.I. (2006) "Pulsed electrodeposition of Zn in the presence of surfactants," *ElectrochimicaActa*, 51, 1342-1350.

[12] T. Hattori, H. Ikezawa, R. Hirano and J. Mochinaga, *Nippon Kagaku Kaishi*, 6 (1982), 952-955.

[13] M. Erdogan (2007)"Tungsten Production by Electrodeoxidation" M.S. thesis, Middle East Technical University.

Rare Metal Technology 2016
Edited by: Shafiq Alam, Hojong Kim, Neale R. Neelameggham, Takanari Ouchi, and Harald Oosterhof
TMS (The Minerals, Metals & Materials Society), 2016

KINETICS OF EXTRACTING VANADIUM FROM STONECOAL BY ALKALI LEACHING

Shengfan Zhou, Bianfang Chen, Mingyu Wang*, Xuewen Wang

School of Metallurgy and Environment, Central South University,
Changsha, Hunan, 410083, China

Keywords: stone coal; vanadium; NaOH; leaching; kinetics

Abstract

Vanadium was leached from roasted stone coal by NaOH solution, and the effect of leaching temperature, leaching time, addition of NaOH, liquid to solid ratio, the grinding fineness of stone coal, and stirring speed on the vanadium leaching rate was studied. The kinetics analysis for the process is presented by the shrinking core model. The results show that the vanadium leaching ratio of roasted stone coal can be beyond 75% at the 6% NaOH addition, leaching temperature of 95°C, leaching time of 7h, liquid to solid ratio of 1.2mL/g, and grinding fineness of 0.111mm. The leaching process is controlled by interfacial reaction kinetics and the apparent activation energy was 59.36kJ/mol.

Introduction

In China, vanadium resources mainly exist in stone coal and vanadium–titanium magnetite. Due to the rapid growth in demand for vanadium products, extraction of vanadium from stone coal has attracted increasing attention.

The traditional process for vanadium extraction from stone coal is roasting with sodium chloride. However, the released exhaust gas containing Cl_2 and HCl is a serious threat to the environment [1]. In recent years, several other processes have been proposed to recover vanadium from stone coal, such as direct leaching with sulfuric acid [2], villiaumite leaching with sulfuric acid [3], microwave assisted sulfation roasting [4], pressure leaching with sulfuric acid [5]. However, due to the low grade of vanadium in stone coal and complex mineral structure, the application of these processes may result in poor recovery and high cost.

Blank roasting-alkali leaching technology characterized by higher recovery of vanadium and environmentally-friendly has been paid widely attention [6,7]. However, the studies of alkali leaching process are not systematic and deep. In this study, the influence factors on vanadium leaching process and leaching kinetics of vanadium were studied to provide a reference for industrialized utilizing of stone coal.

* Corresponding author: Mingyu Wang. Email: wmydxx@163.com

Experimental

Materials and Analysis

The stone coal used in this study was obtained from Huaihua, Hunan province in China. Before leaching, the stone coal was oxidizing roasted at 850℃ for 4h. The chemical composition of the oxidized stone coal is shown in Table 1. Analytic grade NaOH was used as the leaching agent. All the other reagents used in the stone coal analytical process are also analytic grade. The content of vanadium in both initial oxidized stone coal and leaching residue as well as leaching solution was measured by a titration method with ammonium ferrous sulfate.

Table 1 Chemical composition of the roasted stone coal

V_2O_5	MgO	Al_2O_3	SiO_2	K_2O	Na_2O	CaO	Fe	P
1.31	0.43	7.81	80.12	2.78	0.19	0.55	3.13	0.49

Methods

Before leaching, the oxidized stone coal was crushed and grinded. A 1000ml three-neck flask with stirrer was employed to leach the roasted samples in the water bath. The leaching agent NaOH solution was prepared, and mixed with roasted stone coal once at expected temperature in the flask.

In the effect of leaching temperature and time on leaching rate of vanadium research part, the roasted stone coal were leached by NaOH solution with the liquid-solid ratio of 1.5:1 mL/g. Once the selected time intervals reached, 5ml slurry were withdrawn and separated immediately, and then the vanadium concentration in the solution was analyzed. The vanadium leaching rate was calculated based on the vanadium concentration in solution. In other research parts, the vanadium leaching rate was calculated based on the vanadium content in leaching residue.

Results and discussion

Effect of NaOH addition

In this step, the NaOH addition was adjusted between 4% and 12%. The roasted stone coal was leached at 95°C for 7h with a liquid/solid ratio of 1.2 mL/g and stirring speed of 300rpm. The leaching rate of vanadium versus NaOH addition is shown in Fig. 1.

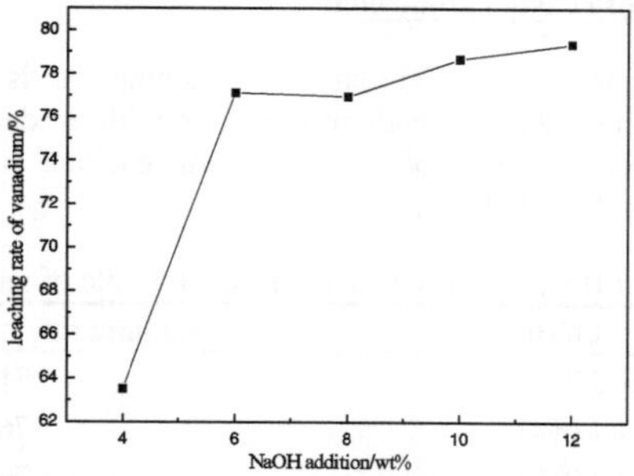

Fig.1 Effect of NaOH addition on vanadium leaching rate

As shown, the vanadium leaching rate increased rapidly to above 77% with the addition of NaOH at 6%. Further addition of NaOH resulted in insignificant increases in the leaching of vanadium. Hence, 6% NaOH addition was chosen as the optimized amount.

Effect of liquid/solid ratio

The effect of the liquid/solid ratio on vanadium leaching was examined with 6% NaOH addition at 95°C for 7 h of leaching. The results obtained are presented in Table 2.

Table 2 Effect of liquid/solid ratio on vanadium leaching rate

L/S(mL/g)	leaching rate of vanadium(%)
0.8	79.59
1.0	80.32
1.2	77.58
1.4	74.85
1.6	75.04

It can be seen from Table 2 that the vanadium leaching rate is close when the liquid/solid ratio is 0.8 mL/g and 1.0 mL/g. After that, the leaching rate of vanadium decreases. However, with the further increase in liquid/solid ratio from 1.4 mL/g to 1.6 mL/g , the vanadium leaching rate become close again. Under the condition that NaOH mass is invariable, the higher liquid/solid ratio, the lower the concentration of OH- in the solution. However, with the increase of liquid/solid ratio, the slurry viscosity reduced which can improve the diffusion conditions to promote the leaching of vanadium. Although a low liquid/solid ratio was favorable to achieve a high concentration of vanadium solution, it would be a disadvantage for agitating and pipeline delivery in an industrial setting. Therefore, the liquid/solid ratio of 1.2 mL/g was chosen.

Effect of particle size of roasted stone coal

The relation between particle size and vanadium leaching rate is shown in Table 3. As can be seen, the leaching rate of vanadium increases with an decrease of particle size, while there's no evident increase of the vanadium leaching rate when particle size continued decreasing under 0.111mm.

Table 3 Effect of particle size on leaching rate of vanadium

particle size (mm)	leaching rate of vanadium (%)
≤0.200 (≥88.7%)	71.93
≤0.111 (≥84.6%)	76.23
≤0.088 (≥82.0%)	76.43

The reaction of stone coal and NaOH solution is a solid-liquid reaction process, and the smaller particle increases the reaction contact area which can accelerate the reaction process. However, the smaller particle has a bigger surface energy, under the condition of a low liquid/solid ratio, thin particles will gather with each other which leads to a lowered effective area. And therefore, when the other conditions are fixed, the effect of particle size on the vanadium leaching rate is not significant anymore after the roasted stone coal was grinded to a certain particle size.

Effect of stirring speed

The relation between the stirring speed and the vanadium leaching rate is shown in Table 4 with the condition of 6% NaOH addition, liquid-solid ratio of 1.2 mL/g, leaching temperature of 95°C, leaching time 7h and the roasted stone coal particle size of 0.111mm.

Table 4 Effect of stirring speed on leaching rate of vanadium

stirring speed (r/min)	leaching rate of vanadium (%)
200	76.97
300	81.33
400	79.75
500	80.07

It can been seen from Table 4 that the vanadium leaching rate of slightly increases with an increase of stirring speed, while the effect of stirring speed on the leaching rate of vanadium is not significant anymore when it was up to 300rpm.

Effect of leaching temperature and time on leaching rate of vanadium

Fig. 2 shows the relationship between the leaching time and vanadium leaching rate. It is clearly seen that the vanadium leaching rate increases with a longer leaching time, but after the leaching time reaches about 450min at a leaching temperature of 60°C and about 400min at 95°C, the vanadium leaching rate rises slowly. When the temperature is elevated the vanadium leaching rate increases remarkably.

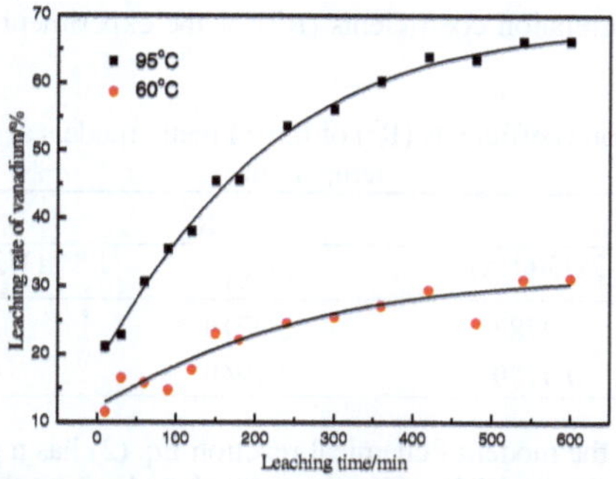

Fig.2 Effect of temperature and time on leaching rate of vanadium

Kinetics analysis

The reaction of roasted stone coal with NaOH is a kind of complex liquid-solid heterogeneous reaction. The kinetics analysis is taking V2O5 as an object of the study. The classic shrinking core model (SCM) considers that interfacial reaction, external diffusion and internal diffusion are the key steps in the leaching process and the leaching speed is decided by the slowest step [8].

Under the condition of a high stirring speed, the effect of external diffusion is negligible. In this research, vanadium leaching was independent over the speed of 200rpm. As a result, the 300 rpm stirring speed is applied to ignore the effect of external diffusion. Assuming the diffusion of NaOH molecules through production layer is the controlling step, the following equation can be used to express the leaching kinetics [9]:

$$1-2x/3-(1-x)^{2/3}=k_1t \quad (1)$$

When the interfacial reaction is the slowest step of the leaching process, the following equation can be used to express the leaching kinetics [10]:

$$1-(1-x)^{1/3}=k_2t \quad (2)$$

where: x is the leaching rate of vanadium, %; t is leaching time, h; k_1 and k_2 are the rate constants.

The Eq. (1) and Eq. (2) only can be applied to describe leaching process when there is one controlled step in the whole process.

When the leaching process was controlled by both the internal diffusion and the diffusion of NaOH through production layer at the same time, the Eq.3 can be presented to describe the kinetics process [11].

$$(1/3)\ln(1-x)+(1-x)^{-1/3}-1=k_3t \quad (3)$$

where: k_3 are the rate constants.

Table 5 shows the correlation coefficients (R2) of the experiment data with these three models.

Table 5 Correlation coefficients (R^2) of three kinetic models at different leaching temperature

T/℃	R^2		
	$1-2x/3-(1-x)^{2/3}$	$1-(1-x)^{1/3}$	$(1/3)\ln(1-x)+(1-x)^{-1/3}-1$
60	0.9787	0.9719	0.9776
95	0.9779	0.9969	0.9279

As shown in Table 5, the model of chemical reaction Eq. (2) has a good fit than the other two models. As a result, model Eq. (2) was applied to describe the kinetics of leaching vanadium from roasted stone coal by sodium hydroxide. Based on the Arrhenius equation [12], the apparent activation energy Ea is calculated as 59.36KJ/mol, which also indicates that the leaching process is controlled by interfacial reaction kinetics.

Conclusions

(1) Vanadium can be leached from roasted stone coal by NaOH solution. The leaching rate of vanadium can reach beyond 75% under the condition of 6% NaOH addition, leaching temperature of 95°C, leaching time of 7h, liquid to solid ratio of 1.2 mL/g, grinding fineness less than 0.111 mm and agitation speed of 300r/min.

(2) The leaching kinetics analysis indicates that the interfacial reaction kinetics affect the leaching process. The calculation apparent activation energy of the leaching process is 59.36KJ/mol.

ACKNOWLEDGMENTS

This study was financially supported by the National Natural Science Foundation of China (No. 51104186) and the China National Scholarship Fund.

References

1. X.W. Wang and M.Y. Wang, "Research Status and Prospect of Vanadium Leaching Processes from Stone Coal", *Chinese Journal of Rare Metals*, 34(1) (2010),90-97.
2. Z.L. Lu, "Investigation and Industrial Practice on Extraction of V_2O_5 from Stone Coal Containing Vanadium by Acid Process", *Hydrometallurgy of China*, 21(4) (2002), 176-182.
3. X.Y. Zhou et al., "Leaching of Vanadium from Carbonaceous Shale", *Hydrometallurgy*, 99 (2009) 97–99.

4. M.Y. Wang et al., "Extraction of Vanadium from Stone Coal by Microwave Assisted *Sulfation Roasting", JOM, 67(2)(2015),369-374.*
5. Z.G. Deng et al.," Extracting Vanadium from Stone-Coal by Oxygen Pressure Acid Leaching and Solvent Extraction," *Transactions of Nonferrous Metals Society of China*, 20(2010), s118-s122.
6. D.S. He et al.,"An Environmentally-friendly Technology of Vanadium Extraction from Stone Coal," *Minerals Engineering* 20 (2007), 1184-1186.
7. A.H. Tan, "Study on Extracting V_2O_5 by the Roasting-alkaline Leaching from Vanadium-containing Stone Coal", *Hunan Nonferrous Metals*, 24(1) (2008), 24-26.
8. T.J. Hu, G.M. Zeng, and X.Z. Yuan, "Leaching kinetics of silver extracted by thiourea from residue in hydrometallurgy of zinc", *The Chinese Journal Of Nonferrous Metals*, 11 (5) (2010), 933-937.
9. Levenspiel,O., *Chemical reaction engineering*.(Wiley New York etc.)
10. X.L. Hua and Z.T. Sui, "Kinetics of leaching Ti-bearing slag by dilute sulphuric acid," *Acta Metallurgica Sinica*, 39 (3) (2003), 293-296.
11. Dickinson. C. and Heal. G., "Solid–liquid diffusion controlled rate equations", *Thermochimica Acta,* 340 (1999.), 89-103.
12. M.H. Wang, X.H. Du. and Z.T. Sui, "Kinetics of acidolysis of rich titanium concentrate by H_2SO_4," *The Chinese Journal of Nonferrous Metals,* 11 (1) (2001), 131-134.

Rare Metal Technology 2016
Edited by: Shafiq Alam, Hojong Kim, Neale R. Neelameggham, Takanari Ouchi, and Harald Oosterhof
TMS (The Minerals, Metals & Materials Society), 2016

SELECTIVE REMOVAL OF THE IMPURITY SILICON AND ALUMINUM IN TITANIUM CONCENTRATE

Xuehui Liu, Xuewen Wang*, Mingyu Wang, Xingming Wang

School of Metallurgy and Environment, Central South University,
Changsha, Hunan, 410083, China

Keywords: Titanium concentrate; Removal impurity; Sodium roasting; Acid leaching

Abstract

The Panzhihua titanium concentrate contains high contents impurity of Mg and Al, and it is difficult to prepare high grade titanium slag. To solve this problem, a sodium salt roasting-acid leaching process was proposed for removing the impurities in titanium concentrate. Na_2CO_3 and titanium concentrate were mixed with the proportion of 1:4, and then the mixture was roasted at 700℃ for 2h. After roasting, the calcine was leached with HCl solution. The experimental results showed that the removal rate of silicon and aluminum can reach more than 85% and 80%, respectively, at leaching temperature of 70℃, leaching time of 4h, HCl addition of 5ml and liquid-solid ratio of 4:1.

Introduction

In nature, ilmenite can be divided into two categories, sand ore and rock ore. The grade (TiO_2) of titanium concentrate came from sands ore are 50% - 64%, while the grade of titanium concentrate obtained from the rock ore usually spread from 42% to 48%. China has rich titanium resources while they are mainly rock ore. Sichuan panzhihua area is the largest vanadium titanium magnetite mining area in the world, its possible ore is above 20 billion tons with 870 million tons symbiotic titanium resource, accounting for 90.54% of the total reserves of titanium resources in China and 35.17% of total world titanium resources reserves [1].

The TiO_2 grade of Panzhihua titanium concentrate is usually less than 47%, duo to it contains a large amount of impurity, such as MgO, CaO, SiO_2 and Al_2O_3[2]. As a result, the TiO_2 grade of the high titanium slag produced form Panzhihua titanium concentrate is too low to use as the materials of chloride method of titanium production. At present, Panzhihua titanium concentrate can only be used for production of titanium white with sulfuric acid method.

In order to economic and efficient use of Panzhihua titanium concentrate, a lot of research has carried out to make synthetic rutile, such as hydrochloric acid leaching method[3], titanium slag-titanium concentrate hydrochloric acid leaching method[4] and sulfuric acid leaching method[5]. All the acid leaching process of above methods can remove the FeO, CaO and MgO from titanium concentrate except for SiO_2 which will directly affect the quality of the rich-titanium material.

To remove of the silicon and aluminum, a method of sodium salt roasting-acid leaching process was proposed. The purpose of the present work is to establish the optimum conditions for the removal of silicon and aluminum from Panzhihua titanium concentrate.

Experimental

Materials and Analysis

Titanium concentrate from Pangang group titanium industry co., LTD was used as the raw material. The sample was ground to 200mesh (75 micron) before roasting. The chemical composition was shown in Table 1. As can be seen that the content of Si, Al, Mg, Ca is high.

Table1 Chemical composition of Panzhihua titanium concentrate, %

Composition	TiO_2	FeO	SiO_2	MgO	CaO	Al_2O_3	other
Raw material	44.6	37.1	7.5	5.8	1.5	2.5	1.0

Figure 1 is the XRD patterns of Panzhihua titanium concentrate. It can be seen from Fig.1 that the mineral compositions of the titanium concentrate are ilmenite($FeTiO_3$), magnesian[$(Fe,Mg)(Ti,Fe)O_3$], idocrase [$Ca_{19}MgFe_2Al_{10}Si_{18}O_{70}(OH)_8$], and andradite [$Ca_3Fe_3(SiO_4)_3$].

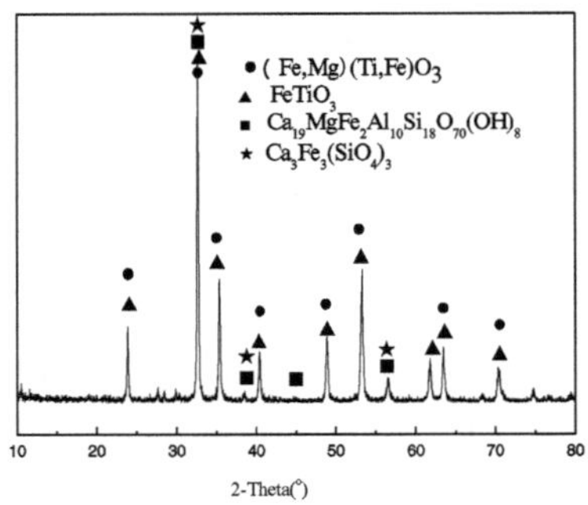

Fig.1 XRD patterns of Panzhihua titanium concentrate

Na_2CO_3, Na_2SO_4, NaCl and hydrochloric acid (12mol/L) used in the experiments are all analytical reagent, the water used in the experiments is deionized water.

Molybdisilicic colorimetry is used to analysis the content of Si, solid sample first handle by alkali fusion, then leach it with boiled water in PTFE beaker, take its filtrate and add ammonium molybdate to color it and measure silica content through colorimetry[5]. Fe, Al, Ca, and Mg in a sample were measured by ICP.

Experimental principles

As shown in Fig.1, Si in titanium concentrate mainly exists in the idocrase and andradite, whose structure is stable and difficult to be destroyed and leached. Roasting with the sodium salt can destroy the lattice construct of idocrase and andradite, and Si will transform into the silicate, which can be removed by acid-leaching process. At the same time, the breaking lattice construct will release other impurities, such as Al, make it easy to be leached, the main reaction equations are as follows:

$$(SiO_2)_c+Na_2CO_3=Na_2SiO_3+CO_2\uparrow \qquad (1)$$

$$(Al_2O_3)_c+6HCl=2AlCl_3+3H_2O \qquad (2)$$

Experimental procedure

Prior to roasting, 40g ground titanium concentrate was placed in a mixing drum, and then uniformly mixed with desired sodium salt. The mixed sample was then transferred to a graphite crucible, and placed in the muffle furnace for roasting at predetermined temperature. After roasting, the calcine was leached with hydrochloric acid solution in an agitated flask while being heated by an electric jacket for designed amount of time. The removal rate of silicon and aluminum was calculated by the following equations:

$$\text{Removal rate} = 100\% - \frac{\text{Weight of Si(Al)in leach residue}}{\text{Weight of Si(Al)in concentrate}} \times 100\%$$

Results and discussion

Effect of additives type

40g titanium concentrate mixed with 10g different kinds of sodium salt and roasting at 700℃ for 4h, then leached by 1% hydrochloric acid with L/S ratio 4:1 at 100℃ for 4h. The relation between the removal rate and the type of the additives is shown in Table 2.

Table 2 Removal rate with different types of the additives, %

Additive	Na_2CO_3	Na_2SO_4	NaCl
Si	78.5	21.1	10.1
Al	84.3	53.4	60.3

As can be seen from Table 2, the addition of Na_2CO_3 has a more significant effect on removal Si and Al than Na_2SO_4 and NaCl, and therefore, the following experiments adopt Na_2CO_3 as a roasting additive.

Effect of roasting temperature on the rate of removal impurity

Fig. 2 is the experimental results of the effect of roasting temperature on the removal rate, which was obtained by adding 10g Na_2CO_3 into 40g titanium concentrate, mixed and roasted at different temperatures for 2h.

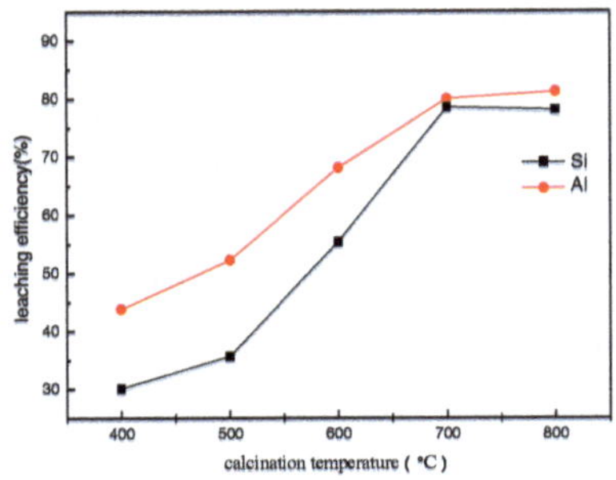

Fig. 2 effect of roasting temperature on the rate of removal impurity

It can be seen from Fig. 2 that the roasting temperature has a significant effect on the removal of impurities of Si and Al. The removal rate of silica and alumina are all increased with the increase of roasting temperature in the range of 400-700. Further increase of roasting temperature resulted in insignification increases in the removal of silica and alumina. And therefore, the roasting temperature of 700℃ was chosen as the optimized temperature.

Effect of roasting time

Fig.3 is the effect of roasting time on the removal rate, which was obtained by adding 10g Na_2CO_3 into 40g titanium concentrate and roasted at 700℃ for different time. The calcine was leached by 1% HCl solution with L/S ratio of 4:1 at 100℃for 4h.

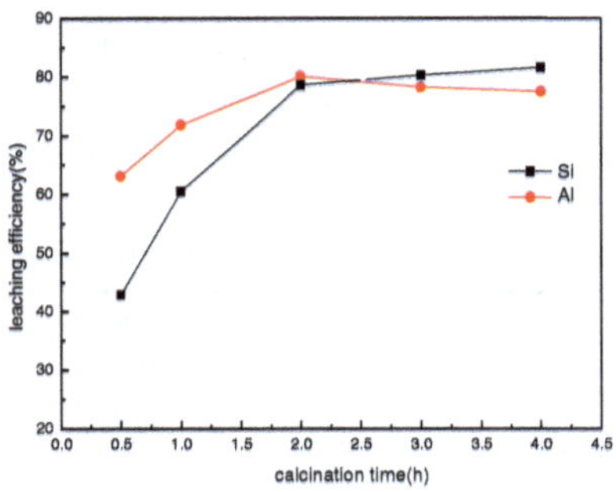

Fig. 3 Effect of roasting time on the removal rate

It can be seen from Fig. 3 that the roasting time almost has no impact on the removal rate of impurities of Si and Al after 2h, that is to say, it's enough long for reacting completely after 2 hours., so choose 2h as the best roasting time.

Effect of Na_2CO_3 addition

Fig. 4 is the effect of Na_2CO_3 addition on the removal rate. The mixed materials was roasted at 700℃ for 2h, and the calcine was leached by 1% hydrochloric acid with L/S ratio of 4:1 at 100℃ for 4h.

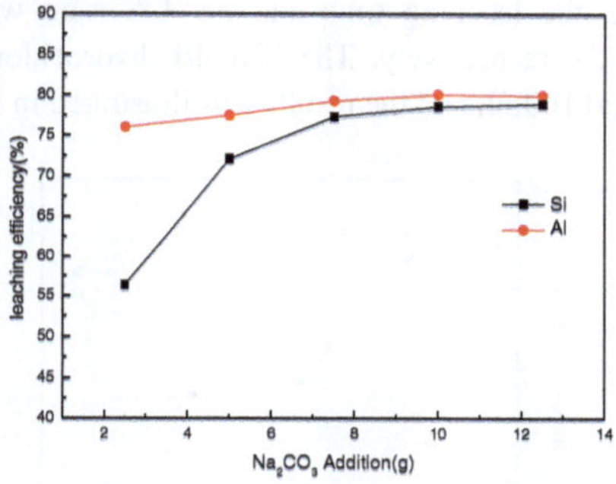

Fig. 4 Effect of Na_2CO_3 addition

As can be seen, the Si removing increased rapidly with the addition of Na_2CO_3. Nevertheless, the same increase was not so clear for Al removing. When the addition of Na_2CO_3 was 10g, the removing of Al and Si can be achieved 80.4% and 78.6%, respectively. Further increase of Na_2CO_3 resulted in insignificant increases in the removing of Al and Si.

Effect of leaching time

Fig. 5 is the effect of leaching time on the removal rate. The calcine was obtained by adding 10g Na_2CO_3 into 40g titanium concentrate and roasted at 700℃ for 2h.

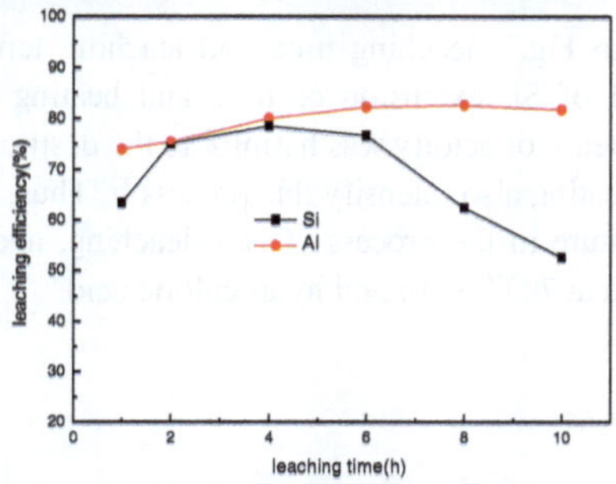

Fig. 5 Effect of leaching time

It can be seen from Fig. 5 that the removal rate of Al has change a little with the time extension. while the removal rate of Si first reached the highest at 4th hour, and then fell down. The reason is that under acid condition, silicic acid will condensation and form colloid precipitation, indeed reduce the leaching rate[6]. Taken together, 4h is appropriate leaching time .

Effect of hydrochloric acid addition

The effect of hydrochloric acid addition on Al and Si removing was studied while the leaching temperature, the leaching time and the L/S ratio were kept constant as 100 °C, 2h and 4 ml/g, respectively. The 12mol/L hydrochloric acid addition was adjusted between 0 and 100ml, and the results was illustrated in Fig. 6.

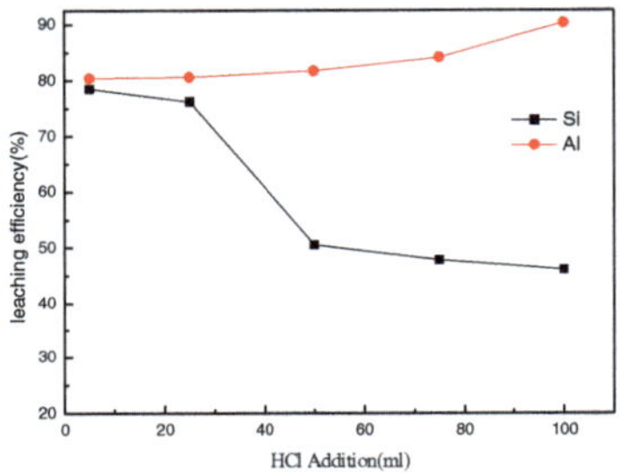

Fig. 6 Effect of leaching acidity on the removal rate

As can be seen from Fig. 6, acidity rise is good for Al removing, and adverse to the removing of Si, the reasons also associated with silicic acid auto-agglutination, rising acidity can also increase the speed of the condensation[7]. However, considering the leaching rate of Al and other metal impurities and reduce the acid consumption and recycling waste acid, with 5ml hydrochloric acid addition is suitable.

Effect of leaching temperature

As seen from Fig.5 to Fig.7, leaching time and leaching temperature have a little influence on removal of Si, extension of time and heating are not conducive to removal of Si, the increase of acidity was harmful to the desilication for the formation of silicic acid while heating also intensify this process[8]. Thus, it is better to keep low acid and low temperature in the process of acid leaching, and the leaching process should carry out for 4h at 70℃ with 5ml hydrochloric acid.

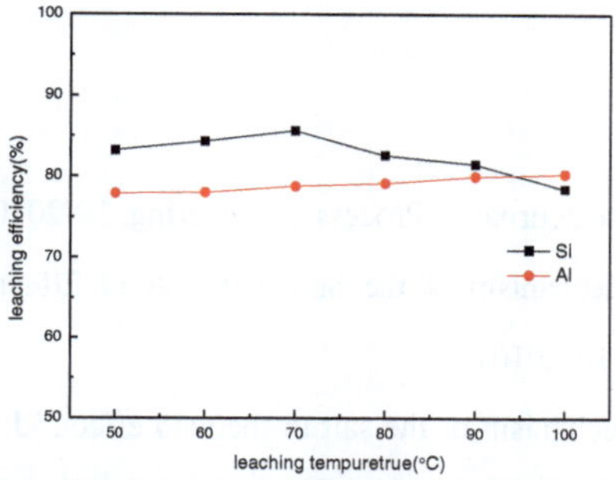

Fig.7 Influence of temperature on the leaching rate

Conclusion

It is an effective technique to remove the silicon and aluminum from Panzhihua titanium concentrate by sodium salt roasting-acid leaching process. Na_2CO_3 and titanium concentrate were mixed with the proportion of 1:4, and roasted at 700℃ for 2h, then the calcine was leached with 5ml HCl solution at 70℃, leaching time of 4h with a liquid-solid ratio of 4:1, under this condition, the removal rate of aluminum can reach more than 80% and the desilication rate can reach more than 85%.

References

1. G. Z. Den, *Metallurgy of Titanium* (Beijing, BJ: Metallurgical Industry Publishing Company, 2010), 56-57.

2. R. Yang, "The Development of Titanium Industry in China in the Nearly Decade," (Paper presented at the China Nonferrous Metals Industry Association Titanium Zirconium and Hafnium Branch Annual Meeting, Dalian,2014).

3. L. SUI, Y. ZHAI, "Reaction kinetics of roasting high-titanium slagwithconcentrated sulfuric acid" *Transactions of Nonferrous Metals Society of China,* 24(2014),848-853.

4. B. X.Yang, "Artificial Rutile Production Technology Development and Technology Selection of Panzhihua Titanium Concentrate Upgrade" (Paper presented at the China Nonferrous Metals Industry Association Titanium Zirconium and Hafnium Branch Annual Meeting, Dalian,2014).

5. Y. J. Liang, C. M. Che, *Inorganic Thermodynamic Data Manual* (Shengyang, SY: Northeastern University Press, 1994)

6. X. B. Liang, "Effect of Preoxidation on Hydrochloric Acid Leaching of Panzhihua

Ilmenite," The Chinese Journal of Process Engineering, 10(2010),

7. D. M. Ji, "The Mechanism of the Salt in the Acid Effect," Chinese Journal of Inorganic Chemistry, 10(2010)

8. Y. F. Guo, "The mechanism of the salt in the acid effect," Journal of University of Science and Technology Beijing, 32(2010)

Rare Metal Technology 2016
Edited by: Shafiq Alam, Hojong Kim, Neale R. Neelameggham, Takanari Ouchi, and Harald Oosterhof
TMS (The Minerals, Metals & Materials Society), 2016

STRIPPING OF Fe(III) FROM P204 BY OXALIC ACID

Changjun Jiang, Shengfan Zhou, Mingyu Wang*, Xuewen Wang

School of Metallurgy and Environment, Central South University,
Changsha, Hunan, 410083, China

Keywords: Oxalic acid, P204, Stripping

Abstract

In the process of extracting vanadium from stone coal by sulfuric acid leaching - P204 solvent extraction, Fe(III) can accumulate in the organic phase which reduces the extraction of vanadium. In order to eliminate the effects of the accumulation of iron, oxalic acid was used as the stripping agent to strip the Fe(III). The effects of oxalic acid concentration, phase ratio O/A, stripping time and shake velocity on the stripping of Fe(III) were studied. The results showed that the Fe(III) stripping rate can reach beyond 96% at the room temperature with 70g/L oxalic acid using an O/A of 2 at 160 rpm shake velocity for 20min. After 2-stage stripping, the concentration of Fe(III) in organic phase was 0.008g/L.

Introduction

Sulfuric acid leaching followed by P204 (D2EHPA) solvent extraction is a typical process of extracting vanadium from stone coal. Industrial practice has proved that when 10%P204+5%TBP+85%sulfonated kerosene was used as the organic phase to concentrate vanadium from the stone coal sulfuric acid leach solution, the extraction process is stable and the two-phase interface of organic phase and solution is clear and easily separated [1,2]. Due to the extraction of Fe(III) is easier than Fe(II), and the extraction of V(IV) and Fe(II) occurs in the order V(IV) > Fe(II), and therefore, the solution containing vanadium was often reduced before extraction [3,4]. However, the residual Fe(III) and the part of oxidized Fe(II) are still extracted by P204 (D2EHPA). 1.5 mol/L H_2SO_4 solution can strip vanadium effectively from the P204, in the meanwhile, Fe(III) remain in the organic phase, thus the purpose of separation of vanadium and iron is fulfilled. After organic phase being used for several times, the accumulated Fe(III) in organic phase will affect its extraction performance. At that time, the Fe(III) must be removed from the organic phase. Usually, 120g/L NH_4HCO_3 solution was used as the stripping agent to remove Fe(III), but it takes a longtime to finish the stripping process and phase separation effect is poor which results in a large loss of the organic phase [5,6].

A drawback of P204 is that the stripping of Fe(III) is very difficult. Even 6mol/L HCl solution was used as the stripping agent, the effect is also not satisfying [7,8]. Liao Y.S. adopt 4~6mol/L NH_4F to strip Fe(III) at 35~40℃ and the single-stage stripping rate can be up to 97%, while this process is not so good for cleaner production [9]. Xiao C. et al.

*Corresponding author: Mingyu Wang, associate professor. Email: wmydxx@163.com.

use sulfuric acid solution to strip the Fe(III) in P204 with iron powders as the reducing agent, the single-stage stripping rate can be up to 73% while it takes 6h to finish the reduction and stripping process [10]. Lupi C et al. adopt zinc powders as reducing agent and dilute sulfuric acid as stripping agent to strip Fe(III) under the condition of vacuum, the single-stage stripping rate can be up to 89% in an hour [11]. Sun J et al. use steel scrap as a reducing agent in the galvanic stripping of iron from P204 and the stripping rate of Fe(III) can be up to 82% after 20 minutes stripping [12]. In this study, the oxalic acid was used as the stripping agent to strip Fe(III), which can provide a new method of the organic phase regeneration in the process of extraction vanadium from stone coal leach solution.

Experimental

Materials

Commercial P204 and sulfonated kerosene (as diluent) were purchased from Shanghai Rare-Earth Chemical Co., Ltd. TBP (tributyl phosphate), ferric sulfate and oxalic acid are analytic grade.10%P204+5%TBP+85% sulfonated kerosene is employed as the organic phase. Other reagents are all analytic grade.

Methods

The stock solution containing 3.1g/L Fe(III) was synthesized by dissolving ferric sulfate in pH=2.0 dilute sulfuric acid. At the room temperature, the Fe(III) solution was fully mixed with the organic phase (10%P204+5%TBP+85% sulfonated kerosene), after extraction, the Fe(III) concentration in organic phase is 2.9g/L. The stripping experiments are all conducted at the room temperature. The effects of oxalic acid concentration, equilibration time, the volume ratio of organic phase to aqueous phase (O/A) and shake velocity on the stripping rate of Fe(III) were studied.

The Fe(III) in aqueous phase was determined by means of o-Phenanthroline spectrophotometry. The Fe(III) content in organic phase was calculated by the subtraction method on the basis of the concentration and volume change of Fe(III) in aqueous phase in the process of extraction and stripping.

Experimental principles

The extraction of Fe(III) can be expressed as follows:

$$Fe^{3+} + 3(HR_2PO_4)_{(o)} = Fe(R_2PO_4)_{3(o)} + 3H^+ \quad (1)$$

Where R is C_8H_{17}.

When oxalic acid was used to strip Fe(III), the complexation reaction of oxalic acid and Fe(III) can make Fe(III) in the organic phase get into aqueous solution. And its reaction equation is as follows:

$$Fe(R_2PO_4)_{3(o)} + 6H^+ + 3C_2O_4^{2-} = H_3Fe(C_2O_4)_3 + 3(HR_2PO_4)_{(o)} \quad (2)$$

Results and discussion

Effect of oxalic acid concentration on iron stripping

The relation between oxalic acid concentration and Fe(III) stripping rate with an O/A of 2 at 160 rpm shake velocity in 20min at room temperature is shown in Fig.1.

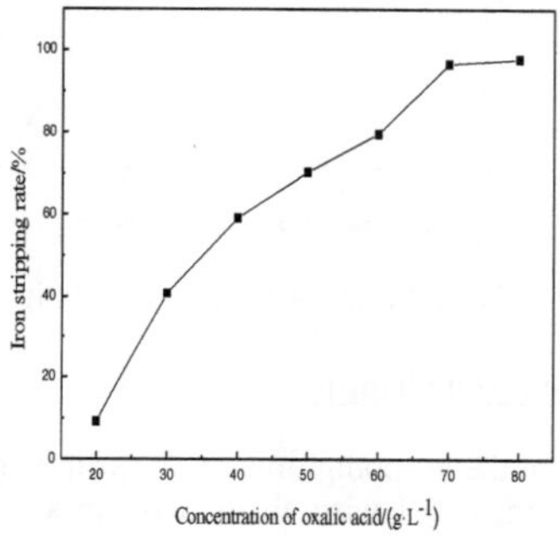

Fig.1 Effect of oxalic acid concentration on iron stripping

It can be seen from Fig.1 that with the increase of oxalic acid concentration, the stripping rate of Fe(III) increased rapidly. When the concentration of oxalic acid increased from 20g/L to 70g/L, the Fe(III) stripping rate increased from 9.2% to 96.7%. Further increase of oxalic acid concentration resulted in insignificant increases in the Fe(III) stripping rate. Hence, 70g/L oxalic acid was selected as the optimal concentration of stripping agent.

Effect of O/A on iron stripping

The effect of O/A on Fe(III) stripping rate was investigated. In this step, the O/A was adjusted between 1:1 and 5:1. The organic phase containing Fe(III) was stripped with 70g/L oxalic acid at 160 rpm shake velocity in 20min at room temperature, and the results are shown in Fig.2.

As can be seen, with the increase of O/A, the Fe(III) stripping rate continually decreased. When the phase ratio O/A was 1:1, the Fe(III) stripping rate was 98.01%. As the O/A was increased to 2:1, the Fe(III) stripping rate was still up to 96.73%. But there's an evident decrease of the Fe(III) stripping rate when the O/A continued increasing. And therefore, the O/A of 2:1 was selected in the following single factor experiment.

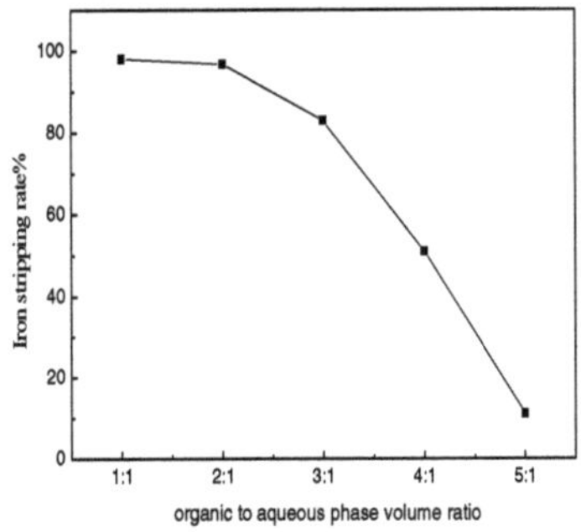

Fig.2 Effect of O/A on iron stripping

Effect of stripping time on iron stripping

Figure 3 shows the effect of the stripping time on the stripping of iron. It is clearly shown that the stripping of iron increases with an increase in stripping time, and after about 20min, the increase trend in iron stripping is no longer obvious with prolonging stripping time. Therefore, the stripping time of 20min was selected in the following single factor experiment.

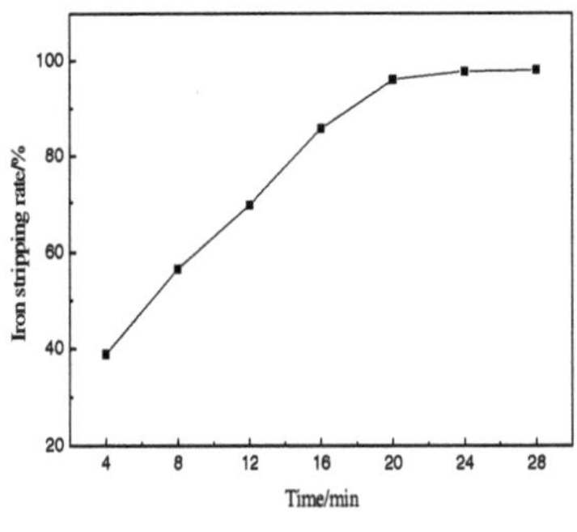

Fig.3 Effect of time on iron stripping

Effect of shake velocity on iron stripping

The relation between shake velocity and Fe(III) stripping rate with 70g/L oxalic acid using an O/A of 2:1 at stripping time of 20min and room temperature is shown in Fig.4.

It can be seen from Fig.4 that the shake velocity has a big impact on Fe(III) stripping rate. When the shake velocity was less than 140rpm, the Fe(III) stripping rate was low, which illustrates that the stripping process is dependent on the mass transport, and at this point the diffusion of the reactants was limited so that the effect of Fe(III) stripping is been impacted. When the shake velocity was increased from 140rpm to 160rpm, the Fe(III) stripping rate sharply rose from 39.5% to 76.1%, and there's no evident increase of the Fe(III) stripping rate when shake velocity continued increasing. Therefore, the shake velocity of 160rpm was selected as the optimal condition.

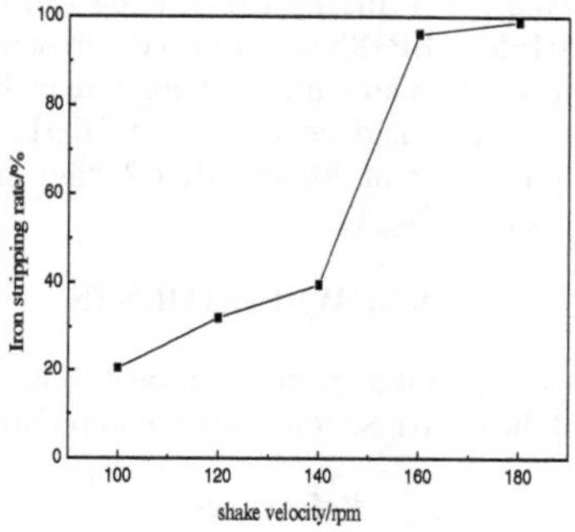

Fig.4 Effect of shake velocity on iron stripping

Stripping isotherm

The result of single factor experiments shows that at the room temperature, the optimal condition of stripping is 70g/L oxalic acid with an O/A of 2 at 160rpm shake velocity in 20min, and under this condition, Fe(III) stripping rate can be up to 96%.

While the organic phase loaded with Fe(III) needed multistage counter-current stripping to get the Fe(III) removed effectively, and therefore, the stripping isotherm was studied. Under the condition of 70g/L oxalic and 160 rpm shake velocity, the Fe(III) stripping distribution isotherm and its McCabe-Thiele diagram are shown in Fig.5.

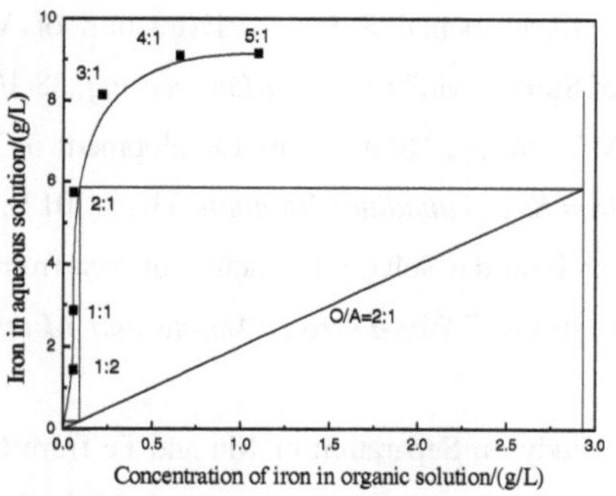

Fig.5 Stripping isotherm of iron

It can be seen that when the O/A is 2:1, in practice, two stripping stages will be good enough. Based on the analysis result, the concentration of Fe(III) in organic phase was 0.008g/L.

Conclusions

With oxalic acid as stripping agent, the Fe(III) in P204 can be stripped effectively. At room temperature,10%P204+5%TBP+85% sulfonated kerosene containing 2.9g/l Fe(III) was stripped by oxalic acid solution and the stripping rate of Fe(III) was more than 96% under optimal conditions: oxalic acid concentration 70g/L, O/A ratio of 2:1, shake velocity of 160 rpm, stripping time of 20min. After 2-stage stripping, the concentration of Fe(III) in organic phase was 0.008g/L.

ACKNOWLEDGMENTS

This study was financially supported by the National Natural Science Foundation of China (No. 51104186) and the China National Scholarship Fund.

References

1. Li S.Y., Xie G. and Yu X.H., "Vanadium Extraction with Acidic Medium," *Nonferrous Metals*, 63(1) (2011), 100-107.
2. Bin Z.Y., "Study on Extraction of V_2O_5 from Vanadium Ore by Roasting and Acid Leaching Process," *Iron Steel Vanadium Titanium*, 27(1) (2006), 22-30.
3. X.B. Li et al., "Solvent Extraction of Vanadium from a Stone Coal Acidic Leach Solution Using D2EHPA/TBP: Continuous Testing", *Hydrometallurgy*, 154 (2015), 40-46.
4. Y.H. Cao et al., "Study on Vanadium Extraction from Vanadium Acid Leaching Solution", *Henan Chemical Industry*, 24(2007), 20-23.
5. X.Y. Chen et al., "Experimental Study on Extraction of Vanadium from Acid Leaching Solution of Stone Coal," *Chemical Engineering*, 38(10) (2010), 146-156.
6. X.W. Wang, and M.Y. Wang, "State-of-art Development of Vanadium Extraction from Stone Coal," *Iron Steel Vanadium Titanium*, 33(1) (2012), 8-14.
7. Ma Y. et al., "Research on the Solvent Extraction of Iron from Acidic Waste Water in Copper Hydrometallurgy," *Non-Ferrous Mining and Metallurgy*, 26 (2) (2010), 24-32.
8. Wang C.Y. et al., "Study on Separation of Mn and Fe from Co Solution by Using P204," *Nonferrous Metals (Extractive Metallurgy)*, 5 (2010), 2-10.

9. Liao Y.S. et al., "Extraction of Vanadium from Stone Coal Leaching Solution," *Nonferrous Metals (Extractive Metallurgy)*, 4 (2009), 13-21.
10. Xiao C. et al., "Reductive Stripping Iron from P204 by Iron Powders and Dilute Sulphuric Acid," *Journal of Central South University (Science and Technology)*, 38 (4) (2007), 663-672.
11. Lupi C and Pilone D, "Reductive stripping in vacuum of Fe(III) from D2EHPA," *Hydrometallurgy*, 57 (2000), 201-208.
12. Sun J. and O'Keefe T.J., "An Evaluation of Steel Scrape as a Reducing Agent in the Galvanic Stripping of Iron from D2EHPA," *Minerals Engineering*, 15 (2002), 177-185.

Rare Metal Technology 2016
Edited by: Shafiq Alam, Hojong Kim, Neale R. Neelameggham, Takanari Ouchi, and Harald Oosterhof
TMS (The Minerals, Metals & Materials Society), 2016

Recovery and Purification of In^{3+} from Zinc Hydrometallurgical Process in a T-junction Microchannel

Chuanhua Li[1,2,4], Feng Jiang [1,2,3], Shaohua Ju[1,3,4*], Jinhui Peng[1,2,3*], Libo Zhang[1,2,3]

1 Yunnan Provincial Key Laboratory of Intensification Metallurgy, Kunming, Yunnan, 650093, China;

2 National Local Joint Laboratory of Engineering Application of Microwave Energy and Equipment Technology, Kunming, Yunnan, 650093, China;

3. Key Laboratory of Unconventional Metallurgy, Ministry of Education, Kunming University of Science and Technology, Kunming, Yunnan 650093, China;

4 Faculty of Metallurgical and Energy Engineering, Kunming University of Science and Technology, Kunming, 650093, China;

Abstract: A T-junction microchannel is proposed to overcome the existing problems in conventional solvent extraction of indium from zinc hydrometallurgical process. Experiments were performed to separate In^{3+}, Fe^{3+} and Zn^{2+} from sulfate solutions using Di-(2-ethylhexyl)phosphoric acid (D2EHPA) as the extractant. The effect of parameters such as pH, microchannel size and contact time on the extraction behavior of In^{3+}, Fe^{3+} and Zn^{2+} was investigated. At contact time of 2.52 s and initial solution pH of 0.48, the extraction efficiency of In^{3+}, Fe^{3+} and Zn^{2+} was 99.13%, 8.31% and 2.45%, respectively and separation factor of $\beta_{In3+/Fe3+}$ and $\beta_{In3+/Zn2+}$ were almost 1229.54 and 4864.98, respectively. Then, the loaded organic phase was stripped in T-junction microchannel. At contact time of 33.6 s, 99.64% of stripping efficiency of In^{3+} and stripping equilibrium were reached, while concentration of Fe^{3+} and Zn^{2+} was lower than 0.01 g/L with good separation of In^{3+} from Fe^{3+} and Zn^{2+}.

Keywords: T-junction microchannel; Solvent extraction; Indium.

---

* **Foundation item:** Project (No.U1302271,No. 51264022 and No.51264015) supported by National Natural Science Foundation of China

Corresponding author: Jin-hui Peng; Tel: +86 8715147924; Fax: +86 8715192076; E-mail: jhpeng@ kmust.edu.cn

Corresponding author: Shaohua Ju; Tel: +8613987662705; Fax: +86 871 65138997; E-mail: shj_200801@ 126.com

1 Introduction

Indium, a rare and valuable metal [1-3], is an important element which is used in the field of electronics [4], energy-related industries [5] and making low melting alloy [6]. Indium has no its own minerals so it has to be recovered as a byproduct from other metallurgical processes [1], together with zinc, lead, copper and tin [6]. Solvent extraction is a commonly used method for the purification of indium from the zinc hydrometallurgical process [4, 7-8]. Thus, the separation efficiency of indium from iron and zinc is of great importance for its purity.

As an emerging new technique, microfluidic technology has many merits such as larger interfacial area and concentration gradient, short diffusion length, facilitating a higher heat and mass rates [9-12]. Zhang Libo [13] et al. studied solvent extraction of Nd(III) in a Y type microchannel. Yang Liqiu [14] et al. investigated copper extraction characteristics in a T-junction microchannel. Maeki Masatoshi [15] used a partially-modified microfluidic device for the extraction behavior of metal ions with calixarene derivatives. The microfluidic solvent extraction can conquer disadvantages such as higher consumption of solvents, low extraction efficiency, significant co-extraction, easily emulsifying phenomenon and fire hazard [16] in conventional solvent extraction so it could be used for recovery of indium.

In the present studies of microfluidic solvent extraction, the chip made by PDMS appears swelling phenomenon and cost is high. The T-junction microchannel made by organic glass won't appear swelling phenomenon that has longer operation life and cost is also lower. Besides, the research of ions is the single and most investigators study two kinds of ions from self-made solution. In addition, there are few reports about stripping experiments in the microchannels.

In this paper, test solution from the zinc hydrometallurgical process was extracted to investigate the extraction behavior of In^{3+}, Fe^{3+} and Zn^{2+} in T-junction microchannel made by organic glass. The effect of parameters such as pH, contact time, microchannel size on extraction efficiency of In^{3+}, Fe^{3+} and Zn^{2+} and the separation factor of In^{3+} against Fe^{3+} were investigated. The loaded organic phase was stripped after Scrubbing and the effect of HCl concentration and contact time on stripping behavior of In^{3+} was also studied.

2 Experimental

2.1 Materials

The test aqueous solution from zinc hydrometallurgy consisting of 3.02 g•L^{-1} of In^{3+}, 4.95 g•L^{-1} of $Fe^{3+}$and 3.57 g•L^{-1} of Zn^{2+} was extracted. The aqueous pH value was adjusted by adding a concentrated sulfuric acid solution (98%) and measured by a pH meter (CHN868, made in China).

The extractant was prepared by dissolving 3 litres D2EHPA solvent into 7 litres 260#oil without saponification operation.

2.2 Apparatus and procedures

The photograph of the microfluidic-extraction system is shown in Figure 1. The organic phase and aqueous solution were respectively pumped into T-junction micro-channel (made by organic glass) from 2 inlets at the T-junction intersection. Then, the two fluids joined into a microchannel where the mixing and extraction reaction took place and then fluids were clarified and separated.

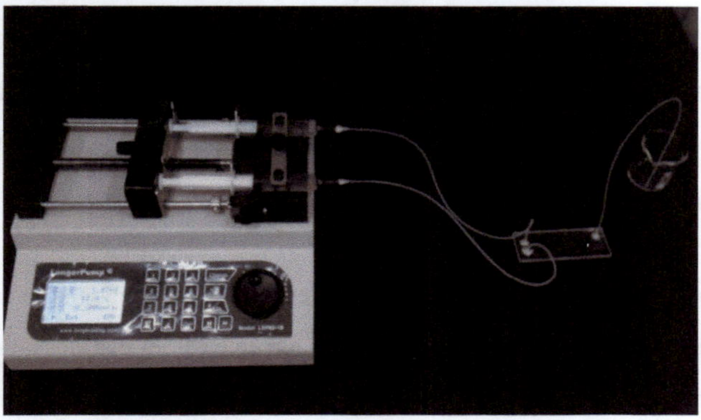

Fig.1. The photograph of the microfluidic-extraction system

The photograph of T-junction micro-channel with length of 140mm is shown in Figure 2.

Fig.2. The photograph of T-junction micro-channel.

2.3 Analytical methods

For all microfluidic and conventional experiments, the raffinate was analysed to estimate the concentrations of In^{3+} and Fe^{3+} using inductively coupled plasma-atomic emission spectroscopy (AA240FS ICP-AES spectrophotometer). The concentration of Zn^{2+} was detected using volumetric method. The concentration in the organic phase was estimated based on mass balance.

2.4 Extraction efficiency and separation coefficient Estimation

According to (Thakur, 1998) [17], in acidic sulfate media, extraction reaction of metal with D2EHPA could be represented by equation of (1.1)

$$(In^{m+})_{aq} + n(HR)_{org} = [InR_m \cdot \frac{n}{\,} HR]_{org} + \frac{n}{\,} H^{+}{}_{aq} \tag{1.1}$$

Contact, t, is calculated from contact of two phases at the beginning of the reaction channel to separation in the outlets, as shown in the following equation:

$$t = \frac{LS}{V} \tag{1.2}$$

In which, L, S, V represent micro channel length, cross-section of micro channel and volume flow rate in micro channel inlets respectively. In this study, the solution was pumped into the different microchips at the same flow rate which resulted in different contact time. In order to facilitate comparison, all flow rates were converted to contact time of all micro channels for investigation.

Extraction efficiency, E, is calculated based on the concentration of metal present as shown in the following equation:

$$E = \frac{C_{aO} - C_{al}}{C_{aO}} \times 100\% \tag{1.3}$$

Where C_{ao} and C_{al} represent metal ion concentration of aqueous phase before and after extraction respectively.

Separation factor, $\beta_{A/B}$, is calculated as shown in the flowing equation:

$$\beta_{A/B} = \frac{D_A}{D_B} = \frac{C_{A,org} \cdot C_{B,aqu}}{C_{A,aqu} \cdot C_{B,org}} \tag{1.4}$$

Farther the departure of separation coefficient from 1 represents better separation, which indicates better the selectivity for the target substance.

Stripping efficiency

$$= \frac{\text{concentration of metal in the aqueous phase after stripping}}{\text{concentration of metal in the organic phase after extraction}} \times 100 \tag{1.5}$$

3. Results and discussion

3.1 Extraction

3.1.1 Effect of pH on the extraction behavior of In^{3+}, Fe^{3+} and Zn^{2+}.

The effect of pH on the extraction behavior of In^{3+}, Fe^{3+} and Zn^{2+} was investigated. These solutions were extracted by 30% v/v D2EHPA in 260# oil at O/A ratio of 1 in the T-junction microchannel with size of 800um and contact time of extraction was 0.336 s.

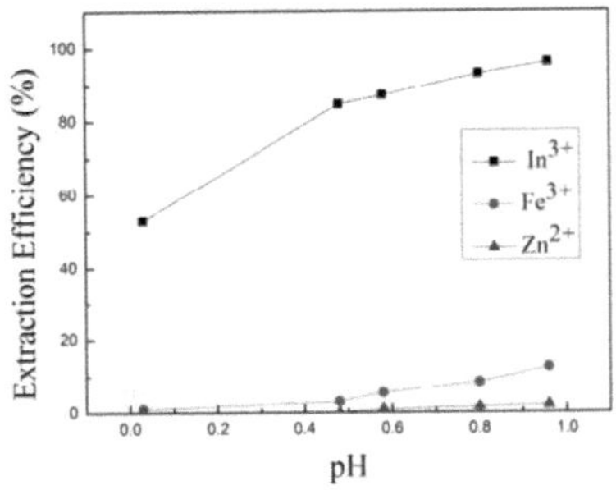

Fig.1. Effect of pH on the extraction efficiency of In^{3+}, Fe^{3+} and Zn^{2+}.

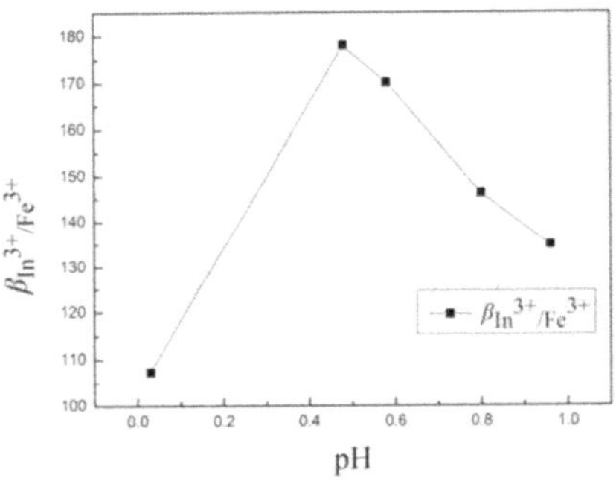

Fig.2. Effect of pH on the Separation factor of $\beta_{In3+/Fe3+}$.

Fig.1 shows the variation of extraction efficiency of In^{3+}, Fe^{3+} and Zn^{2+} against pH and the extraction efficiency of In^{3+}, Fe^{3+} and Zn^{2+} increased with the increase of pH. Extraction reaction of metals with D2EHPA could be represented by (Thakur, 1998) [17] equation of (1.1) so higher pH could strengthen extraction behavior. At pH of 0.48, the extraction efficiency of In^{3+} and Fe^{3+} was 84.67% and 3.01% respectively and Zn^{2+} was not extracted at all from the solution.

At pH of 0.48, the separation factor of $\beta_{In3+/Fe3+}$ was 178.07 and $\beta_{In3+/Fe3+}$ did not exceeded 178.07 with the increase of pH and the separation factor of $\beta_{In3+/Zn2+}$ was thoughtless because Zn^{2+} was not extracted at all from solution. From these experimental results, it is believed that D2EHPA has good selectivity for indium over iron and zinc at pH of 0.48 so other extraction experiments were carried out in this condition.

3.1.2 Effect of contact time on the extraction behavior of In^{3+}, Fe^{3+} and Zn^{2+}.

The effect of contact time on extraction efficiency at the phase ratio of 1:1 in the T-junction microchannel with size of 800um was investigated, and the results are shown in Fig.3.

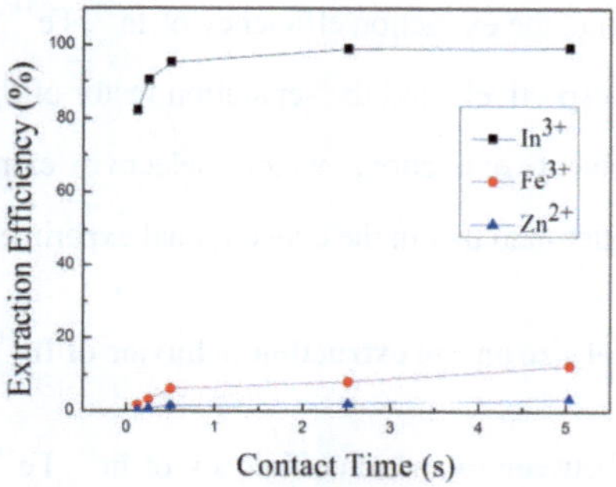

Fig.3. Effect of contact time on the extraction efficiency of In^{3+}, Fe^{3+} and Zn^{2+}.

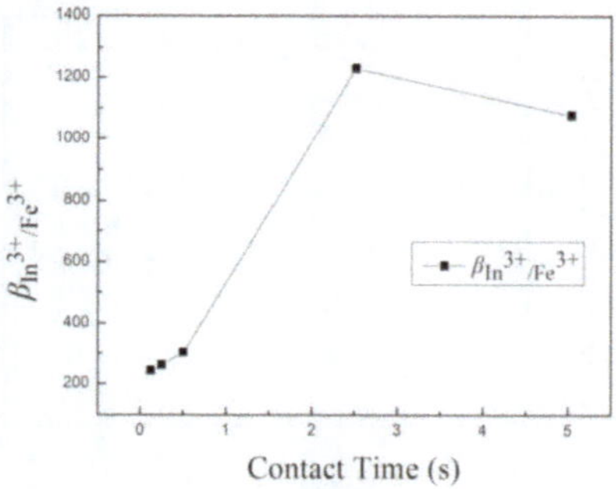

Fig.4. Effect of contact time on the Separation factor of $\beta_{In3+/Fe3+}$.

The extraction equilibrium in T-junction micro-channel was rapid and contact time played an important role in the separation of In^{3+}, Fe^{3+} and Zn^{2+}. Fig.3 and Fig.4 show that the extraction efficiency of In^{3+}, Fe^{3+} and Zn^{2+} and separation factor of $\beta_{In3+/Fe3+}$ increased with the increase of contact time. Except for the case when the contact time was over 2.52 s, the separation factor of $\beta_{In3+/Fe3+}$ began to reduce which indicated the selectivity for indium over iron became bad.

At contact time of 2.52 s, the extraction efficiency of In^{3+}, Fe^{3+} and Zn^{2+} was 99.11%, 8.33% and 2.24% respectively. The separation factor of $\beta_{In3+/Fe3+}$ at contact time of 2.52 s was 1229.54 while that at contact time of 5.04 s was 1077.02 which indicated there was better selectivity for indium over iron at contact time of 2.52 s. A suited contact time should be chosen by metals content of aqueous solution extracted. At contact time of 3 min, the extraction efficiency of In^{3+}, Fe^{3+} and Zn^{2+} was 99.01%, 39.46% and 19.29% respectively and the separation factor of $\beta_{In3+/Fe3+}$ was 428.05 in the conventional solvent extraction, which selective extraction in T-junction micro-channel was better than that in the conventional experiments.

3.1.3 Effect of channel size on the extraction behavior of In^{3+}, Fe^{3+} and Zn^{2+}.

The comparison between extraction efficiency of In^{3+}, Fe^{3+} and Zn^{2+} in different channel size is shown in Fig.4. The test solution was extracted by 30% v/v D2EHPA in 260#oil at O/A ratio of 1 and pH of 0.48.

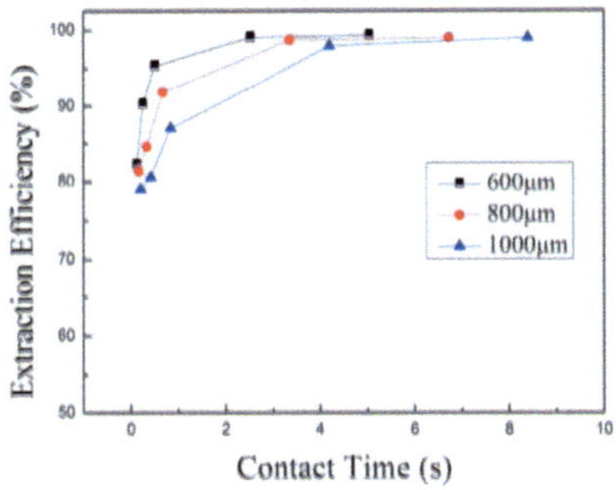

Fig.5. Effect of channel size on the extraction efficiency of In^{3+}, Fe^{3+} and Zn^{2+}.

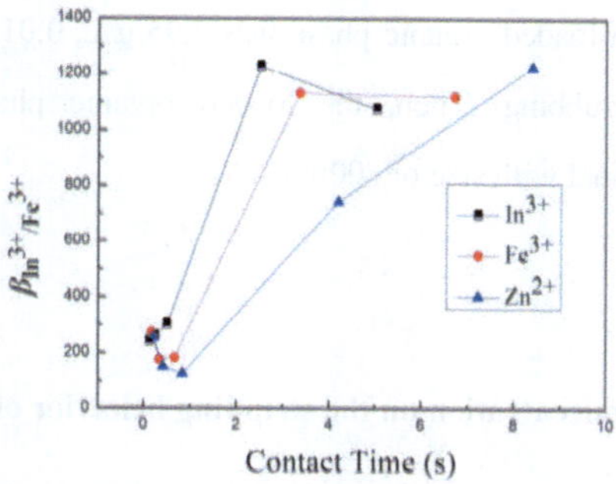

Fig.6. Effect of channel size on the separation factor of $\beta_{In3+/Fe3+}$.

The Fig.5 shows that the smaller the channel size is, the quicker and higher In^{3+} extraction is. When above 99% of the extraction efficiency and extraction equilibrium of In^{3+} were reached, the contact time in channel size of 600um, 800um and 1000um was 2.52s, 6.72s and 8.40s, respectively. The Fig.6 shows that the smaller the channel size is, the higher separation factor of $\beta_{In3+/Fe3+}$ is before extraction equilibrium of In^{3+}.The result reveals that the microfluidic reactor with smaller channel size results in high mass transfer rate, due to its high concentration gradient and short diffusion distance.

3.2 Scrubbing

From the extraction experiments, it was observed that D2EHPA could extract indium selectively from the In–Fe–Zn test solutions. However, small amounts of iron and zinc were still extracted into loaded organic phase. Therefore, scrubbing experiments for farther separation of the metals were conducted. The loaded organic phase was prepared by microfluidic solvent extraction in T-junction micro-channel with size of 800um. The concentration of metal ions of In^{3+}, Fe^{3+} and Zn^{2+} in loaded organic phase was 2.99 g/L, 0.41 g/L and 0.08 g/L, respectively. The loaded organic phase was scrubbed by 180 g/L H_2SO_4 in T-junction micro-channel with size of 800um. At contact time of 0.336 s, most Fe^{3+} and Zn^{2+} were scrubbed away while most In^{3+} was still kept in loaded organic phase. The concentration of metal ions of

In^{3+}, Fe^{3+} and Zn^{2+} in loaded organic phase was 2.95 g/L, 0.012 g/L and 0.005 g/L, respectively after scrubbing. Then, the loaded organic phase was stripped in T-junction micro-channel with size of 800um.

3.3 Stripping

3.3.1 Effect of HCl concentration on the stripping behavior of In^{3+}.

The effect of HCl concentration on stripping behavior of In^{3+} at the phase ratio of 1:1 and contact time of 0.672 s in T-junction micro-channel with size of 800um was investigated, and the results are shown in Fig.7.

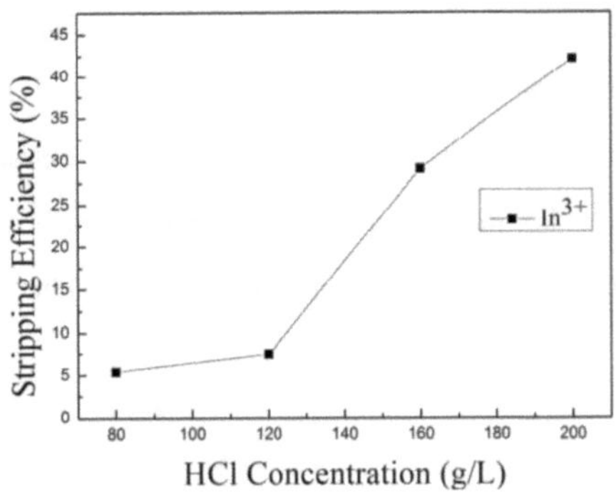

Fig.7. Effect of HCl concentration on the stripping behavior of In^{3+}.

The Fig.7 shows that the stripping efficiency of In^{3+} increased with the increase of HCl concentration in the range from 80 g/L to 200 g/L. Higher hydrochloric acid concentration would corrode the equipment so 160 g/L hydrochloric acid was chosen to strip In^{3+}.

3.3.2 Effect of contact time on the stripping behavior of In^{3+}.

The effect of contact time on stripping behavior of In^{3+} at the phase ratio of 1:1 and 160 g/L hydrochloric acid in T-junction micro-channel with size of 800um was investigated, and the results are shown in Fig.8.

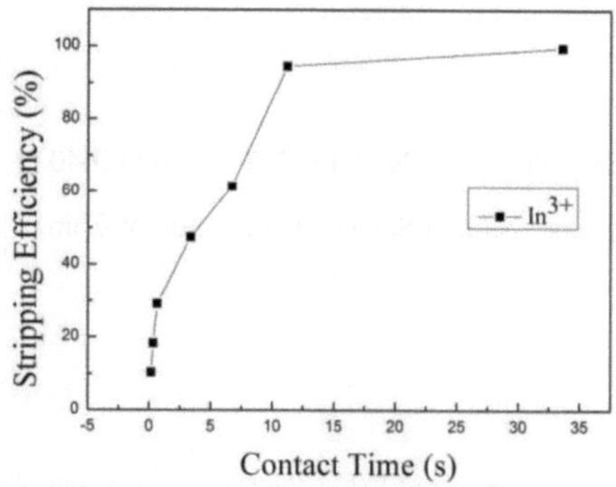

Fig.8. Effect of contact time on the stripping behavior of In^{3+}.

The Fig.8 shows that the stripping efficiency of In^{3+} increased with the increase of contact time. At contact time of 33.6 s, 99.64% of stripping efficiency of In^{3+} and stripping equilibrium were reached while that needed 3.5 min in conventional stripping. Finally, the concentration of In^{3+} was about 2.94 g/L while concentration of Fe^{3+} and Zn^{2+} was lower than 0.003 g/L which indicated a good separation of In^{3+} against Fe^{3+} and Zn^{2+} by extraction, scrubbing and stripping in T-junction micro-channel.

4. Conclusion

The effect of parameters on the extraction behavior of In^{3+}, Fe^{3+} and Zn^{2+}, scrubbing and stripping behavior of In^{3+} from the zinc hydrometallurgical process were investigated in T-junction microchannel. In the optimal conditions, the contact time of extraction, scrubbing and stripping was 3.36s, 0.336s and 33.6s, respectively. The concentration of In^{3+} was about 2.94 g/L while concentration of Fe^{3+} and Zn^{2+} was lower than 0.003 g/L with good separation of In^{3+}, Fe^{3+} and Zn^{2+}. In addition, the extraction efficiency and separation factor were simultaneously proposed to determine the optimal conditions, which make applicability more extensive.

Acknowledgment

Foundation item: Projects (No.U1302271, No.51264022 and No.51264015) supported by the National Natural Science Foundation of China.

References

1. Sami Virolainen, Don Ibana, and Erkki Paatero, "Recovery of indium from indium tin oxide by solvent extraction," *Hydrometallurgy*, 107(2011), 56–61.

2. Birendra Babu Adhikari, Manju Gurung, and Hidetaka Kawakita, "Complexation and extraction behavior of trivalent indium with multiple proton ionizable p-t-butylcalix[5]arene pentacaarboxylic acid derivative: a new efficient solvent extraction reagent for indium," *Journal of Inclusion Phenomena and Macrocyclic Chemistry*, 71 (2011), 479–487

3. Birendra Babu Adhikari, Manju Gurung, and Hidetaka Kawakita, "Multiple proton ionizable calixarene derivatives with different ring sizes and complexation ability as efficient ionophores for complexation and solvent extraction of trivalent indium," *Analyst*, 136 (2011):4570-4579.

4. M.C.B. FORTES, and J. S. BENEDETTO, "Separation of indium and iron by solvent extraction," *Minerals Engineering*, 2 (1998), 447–451.

5. Hee Nam Kang, Jin-Young Lee, and Jong-Young Kim, "Recovery of indium from etching waste by solvent extraction and electrolytic refining," *Hydrometallurgy*, 110(2011), 120–127.

6. S.V. Mahamuni, S.S. Kolekar, and P.P. Wadgaonkara, "Solvent Extraction of Trivalent Indium from Succinate Solution by 2-Octylaminopyridine in Chloroform," *Journal of the Iranian Chemical Society*, 6(2009), 200-212.

7. J.L. Ruan, Y.W. Guo, and Qi Qiao, "Recovery of indium from scrap TFT-LCDs by solvent extraction," *Procedia Environmental Sciences*, 16(2012), 545 – 551.

8. Hee-Nam Kang, Kwan-Young Kim, and Jong-Young Kim, "Recovery and

purification of indium from waste sputtering target by selective solvent extraction of Sn," *Green Chemistry*, 15(2013), 2200–2207.

9. J. H. Xu, G. S. Luo, G. G. Chen, and B. Tan, "Mass transfer performance and two-phase flow characteristic in membrane dispersion mini-extractor," *Journal of Membrane Science*, 249(2005), 75-81.

10. P. Löb, H. Pennemann, V. Hessel, and Y. Men, "Impact of fluid path geometry and operating parameters on l/l-dispersion in interdigital micromixers," *Chemical Engineering Science*, 61 (2006):2959–2967.

11. A. L. Dessimoz, L. Cavin, A. Renken and L. Kiwi-Minsker, "Liquid–liquid two-phase flow patterns and mass transfer characteristics in rectangular glass microreactors," *Chemical Engineering Science*, 63 (2008):4035–4044.

12. Kashid, Madhvanand N., A. Renken, and L. Kiwi-Minsker, "Gas–liquid and liquid–liquid mass transfer in microstructured reactors," *Chemical Engineering Science*, 66 (2011):3876–3897.

13. Zhang Libo, Xie Feng, and Li Shiwei, "Solvent extraction of Nd(III) in a Y type microchannel with 2-ethylhexyl phosphoric acid-2-ethylhexyl este," *Green Processing and Synthesis*, 1(2015):3-10.

14. Yang Liqiu, Zhao Yuchao, and S, Yuanhai, "An Experimental Study of Copper Extraction Characteristics in a T-Junction Microchannel," *Chemical Engineering & Technology*, 6(2013), 985-992.

15. Maeki, Masatoshi, Y. Hatanaka, and K. Yamashita, "Solvent Extraction Behavior of Metal Ions with Calixarene Derivatives by Using Microreactor," *Solvent Extraction Research & Development Japan*, 21(2014), 77-82.

16. Fukiko KUBOTA, Jun-ichi UCHIDA, and Masahiro GOTO, "Extraction and separation of rare earth metals by a microreactor," *Solvent Extraction Research and Development-Japan*, 10(2003), 93–102.

17. Thakur, N. V., "Extraction studies of base metals (Mn, Cu, Co and Ni) using the extractant 2-ethylhexyl 2-ethylhexyl phosphonic acid, PC88A," *Hydrometallurgy*, 48 (1998), 125–131.

AUTHOR INDEX
Rare Metal Technology 2016

SUBJECT INDEX
Rare Metal Technology 2016

MIX
Papier aus verantwortungsvollen Quellen
Paper from responsible sources
FSC® C105338

If you have any concerns about our products,
you can contact us on
ProductSafety@springernature.com

In case Publisher is established outside the EU,
the EU authorized representative is:
Springer Nature Customer Service Center GmbH
Europaplatz 3, 69115 Heidelberg, Germany

Printed by Libri Plureos GmbH
in Hamburg, Germany